Analytical Aspects of Mercury
and Other Heavy Metals in the Environment

CURRENT TOPICS IN ENVIRONMENTAL AND TOXICOLOGICAL
CHEMISTRY

This is a series of books based on papers from the two journals, *International Journal of Environmental and Analytical Chemistry* and *Toxicological and Environmental Chemistry Reviews*.

Volume 1   ANALYTICAL ASPECTS OF MERCURY AND OTHER
HEAVY METALS IN THE ENVIRONMENT
edited by R. W. Frei and O. Hutzinger

Analytical Aspects of Mercury and Other Heavy Metals in the Environment

Edited by

R. W. FREI

Analytical Research and Development,
Pharmaceutical Department,
Sandoz Limited, Basel,
Switzerland

and

O. HUTZINGER

National Research Council, Atlantic
Laboratory, Halifax, Nova Scotia, Canada
also
Present Address: Laboratory of
Environmental Chemistry, University
of Amsterdam, The Netherlands

GORDON AND BREACH SCIENCE PUBLISHERS

LONDON   NEW YORK   PARIS

Copyright © 1975 by

Gordon and Breach Science Publishers Ltd.
42 William IV Street
London W.C.2.

Editorial office for the United States of America

Gordon and Breach Science Publishers, Inc.
One Park Avenue
New York, N.Y. 10016

Editorial office for France

Gordon & Breach
7–9 rue Emile Dubois
Paris 75014

The articles published in this book first appeared in the *International Journal of Environmental Analytical Chemistry*, Volumes 1 and 2, Numbers 1, 2 and Numbers 1, 2 and 4; some were published in *Toxicological and Environmental Chemistry Reviews*, Volume 1, Number 4 and Volume 2, Number 1.

Contents

Preface

This book has been prompted by the increased awareness of the harmful effects of mercury and other heavy metals, such as lead, cadmium, antimony etc., in our everyday life. In the course of these developments limits of contamination have been set by many authorities all over the industrialized world. In order to enforce this legislation and to give meaning to the many data that have been produced, a sound knowledge of the analytical chemistry of mercury and these other metals is essential. Methods by which concentrations of these pollutants can be measured at the very low concentrations at which they are known to be toxic are badly needed.

How meaningful are the results gathered in the past? What are the sampling problems involved? How does one choose an optimal method for a particular problem and fitted best to a particular matrix? These are but a few of the questions for which answers can be found in this book. While it does not claim to be in any way complete in this wide and complex area it will provide up to date reviews, particularly for mercury, and a guide to recent literature. It will give some of the most recent and successful methods with enough detail for a direct adoption in the laboratory.

It is hoped that this book will help to stimulate meaningful research in the area of heavy metals studies and problem-solving on the environmental scene. The editors feel that it should prove useful also to scientists not well acquainted with these pollutants or in areas which have to make use of heavy metal analysis such as Biology, Geology, Oceanography, Medicine Toxicology and Pharmacology.

Use of Mercury in Agriculture and its Relationship to Environmental Pollution

J. G. SAHA and K. S. McKINLAY

Canada Agriculture Research Station, Saskatoon, Saskatchewan, Canada

Contents

I. INTRODUCTION

Man has used cinnabar, the principal ore of mercury, as a pigment since pre-historic times.[1] Mercury compounds have been used by Arab and Greek physicians since the 4th century B.C.[1] and they are still being used effectively and safely to treat a variety of infections and disorders today. The toxic properties of mercury compounds have also been known for a long time and they have been used at times for murder or suicide.[2]

1

However, it is only recently that the toxicity of mercury has caused wide-spread public concern as the result of several widely-publicized disasters.[3-7] These incidents took place in limited areas where there had been very heavy localized pollution with mercurial compounds. The first incident was in 1950 at Minamata, Japan, where 111 people were poisoned after eating sea food containing high levels (27–102 ppm dry weight) of methylmercury.[5,8] Of these people, 22 were injured before birth when their mothers ate contaminated food during pregnancy. When the Minamata report was written in 1968, 41 patients had died of whom two had been poisoned pre-natally. In 1964–65 a similar epidemic occurred in Niigata, Japan, when 26 people were poisoned after eating sea food contaminated with methylmercury compounds and five died.[9] In both of these instances the water had been polluted by effluents from chemical plants.[5,8-10]

Subsequently, commercial fishing was banned on some 40 lakes and rivers in Sweden in 1967 when the fish were found to contain more than 1 ppm of mercury. Similarly, commercial fishing was banned on many lakes and rivers in North America when mercury residues greater than 0.5 ppm were found. The source of mercury in all these cases was pollution by industrial effluents.

Public concern was further aroused by several poisoning incidents which arose from the misuse of mercury compounds in agriculture. In 1956 and 1960 hundreds of people in Iraq were poisoned when they ate bread made from grain treated with mercurial seed dressings. The treated grain had been issued to farmers for planting but economic conditions resulted in its improper use as food. Similar outbreaks were reported later from Guatemala and Pakistan.[2] In January 1970, three children in New Mexico, U.S.A., were poisoned after eating pork from hogs which had been improperly fed treated grain.[11,12]

These incidents certainly show that mercury compounds can be potentially dangerous environmental pollutants under certain conditions. However, this does not mean that all forms of mercury are dangerous in all circumstances, as members of the public have tended to conclude. Nor can one assume that all mercury residues found in plant and animal material are due to man-made pollution. Mercury from natural sources is widely distributed at low levels throughout the environment and it is important to distinguish residues with a natural origin from those due to man's activities.

The agricultural use of mercury has been the origin of considerable concern and although some of the criticism has been justifiable much of it has been without any scientific basis at all. The present review attempts to make a realistic assessment of the risks involved in the agricultural use of mercury. In order to put the problem in perspective, however, one must also consider the natural occurrence of mercury in the environment and the changes brought about by other human activities.

II. MERCURY IN THE ENVIRONMENT FROM NATURAL SOURCES

The over-all position of mercury as an environmental pollutant has been reviewed by Saha[13] and no attempt will be made to discuss the problem in detail here. The essential point to remember is that mercury has certain characteristics which ensure that it will be distributed naturally throughout all parts of the environment. It is the only metal which is liquid at room temperature, and it boils at 357°C, compared to 2212°C for silver for example. It is ten thousand times more volatile than DDT and has a vapour pressure of 1.2×10^{-3} compared to 1.5×10^{-7} mm for DDT.[14] Mercuric sulphide or cinnabar, the only commercial ore of mercury, volatilizes without melting at atmospheric pressure and its vapour pressure reaches 1 atmosphere at only 500°C.[15] The volatility of mercury and its compounds is such that air sampling has been used in the search for ore bodies of this and other related metals.[16,17] In tests carried out by the U.S. Geological Survey[18] air samples were taken from an aircraft flying at an altitude of 200 ft above ground level. Samples taken over areas with known mercury deposits averaged 20 times the general background level. In tests at Cortez, Nevada, samples of air collected at ground level contained anomalous amounts of mercury which correlated with known gold-bearing rocks concealed beneath as much as 100 ft of gravel.

The occurrence of mercury ores is also significant. Cinnabar tends to be found in shallow deposits at depths varying from a few feet to 2000 ft. Most U.S. mines are less than 200 ft deep and only a few are more than 1000 ft.[1] The volatile nature of the material and the shallowness of the deposits have resulted in the worldwide distribution of traces of mercury by aerial transport. Most of this mercury vapour will be washed out of the air by rain and it is estimated[19] that the earth receives 100,000 tons of mercury annually in precipitation. Further mercury will be added to the environment by such processes as leaching and volcanic activity.

It is not surprising that mercury is of almost universal occurrence in small amounts throughout the environment. H. M. Rouelle and J. L. Proust reported mercury in sea-water in 1777 and 1779, respectively.[20] Stock and Cucuel[21] found that rain-water collected at various places and times contained between 0.05 and 0.48 ppb of mercury and averaged 0.2 ppb. Heide *et al.*[22] found that the water of the Saale river contained 0.067 ppb of dissolved mercury and 0.021 ppb of suspended mercury. Hosohara[23] found mercury concentrations in sea-water to increase with depth. He found concentrations of mercury in surface water at four stations over the Ramapo Deep (30°N, 139–48°E) in the Pacific Ocean to be 0.11 ppb and concentrations at a depth of 3000 m were 0.15 to 0.27 ppb.

The earth's crust contains an average of 0.5 ppm of mercury[24] and sedimentary rocks contain more mercury than the igneous rocks.[25] English soils analyzed by Martin[26] contained between 0.01 and 0.06 ppm of mercury. Anderson[27] reported 0.02 to 0.9 ppm mercury in some 200 analyses of Swedish soils. Stock and Cucuel[21] reported up to 1.7 ppm mercury in garden soil. Sand et al.[28] found 0.02 to 2.05 ppm in agricultural soils from seven different states of the U.S.A. where no mercurial seed dressings had been used.

When an element is so widely distributed one would expect most living organisms to contain traces of it. Warren et al.[29] found that when soil contained less than 1 ppm of mercury all plants growing in the soil tended to concentrate it. When soils had very high levels of mercury, tens or hundreds of parts per million, the vegetation contained less mercury than the related soil. In some cases a very high degree of concentration can take place. Rankama and Sahama[25] reported that droplets of metallic mercury can be found in the seed capsules of the jagged chickweed, *Holosteum umbellatum*, growing on mercury-rich soils. According to these authors some species of marine algae can contain as much as 100 times as much mercury as the surrounding sea-water. Predatory fish, such as pike, may have 3000 times the mercury level of the surrounding water.[30]

In these circumstances the discovery of mercury residues in tuna and swordfish is not surprising; in fact, it would be much more remarkable if they were not found! All living creatures on the earth have evolved in the almost universal presence of traces of mercury of natural origin and predatory species have always been exposed to higher levels than the rest. Thus the presence of mercury residues does not necessarily imply man-made pollution, nor is it necessarily disastrous. These residues may have been present ever since life started to evolve on the earth. Whether these mercury residues, either natural or due to man's activities, are toxic is a separate question considered later in the review.

III. MERCURY IN THE ENVIRONMENT DUE TO MAN'S ACTIVITIES

Superimposed upon the natural occurrence and distribution of mercury there is additional concentration and redistribution due to man's activities. In fact, man's contribution is relatively small. As stated before, the earth's surface receives an estimated 100,000 tons of mercury per annum through precipitation[19] compared with a total world production of 10,000 tons per annum.[18] Man's activities cause problems because they create very high levels of mercury in very limited locations. It is this localized pollution which can have disastrous consequences, as was demonstrated in Minnamata and Niigata.

There was little commercial use of mercury until 1557 when Bartoleme de Medina invented the Patio process for the recovery of silver by amalgamation. Since then the production of mercury for commercial use has increased steadily to reach the present production level of about 10,000 tons per annum. In 1968 the world production of mercury was 19.3 million lb and the U.S. consumption was 5.7 million lb.[18] In one estimate, about 72% of all mercury used in the U.S.A. in 1968 was lost to the environment.[2]

The biggest consumer of mercury in North America is the chlor-alkali industry (Table I) which is perhaps the most common source of dangerous,

TABLE I

Consumption of mercury in the U.S. in 1969
(ref. 1)

Uses	Mercury consumed	
	thousands of pounds	% of total
Chlor-alkali plants	1572	26.0
Electrical equipment	1382	22.9
Paint	739	12.2
Instruments	391	6.5
Catalysts	221	3.6
Dental preparations	209	3.5
Agriculture	204	3.4
General laboratory use	126	2.1
Pharmaceuticals	52	0.9
Pulp and paper making	42	0.7
Amalgamation	15	0.2
Other[a]	1082	17.9
Total	6035	99.9

a Includes purchases for expansion and new chlor-alkali plants.

localized, high concentrations of mercury, particularly in water. Sixty-six percent of all mercury used in Canada is consumed by this industry.[7] It has been estimated that about 0.35 to 0.5 lb of mercury can be lost to the environment for every ton of chlorine produced.[3,31] Plants producing 100 tons or more of chlorine per day are quite common and a plant of this size can discharge 9 to 18 thousand lb of mercury into the environment annually, mainly in the effluent to local bodies of water. Although the pulp and paper industries in the U.S.A. use very little mercury, this industry was found to contribute to the pollution of many waters in Sweden.

Numerous studies have shown that fish caught downstream from plants discharging mercury in their effluents have significantly higher levels of

TABLE II

Effect of direct pollution of water with mercury containing industrial waste on residue content of aquatic organisms

Material	Locality	Mercury (ppm)	Reference
Perch	Downstream from a paper mill	1.91–3.48	32
Perch	Upstream from same mill	0.18–0.70	32
Perch	Downstream from a rectifier factory	0.83–2.48	32
Eelpout	Upstream from the same factory	0.35–0.70	32
Pike	Downstream from a pulp mill	1.50–3.00	33
Pike	Upstream from the same mill	0.16–0.83	33
Pike	Upstream from a chlor-alkali plant	0.50–1.70	34
Pike	Downstream from the same plant	6.10–10.60	34
Sauger	Upstream from a chlor-alkali plant	0.42–0.84	35
Pike	Downstream from the same plant	2.20–6.11	35
Goldeye	Downstream from the same plant	0.96–4.25	35
Pike	Upstream from a chlor-alkali plant	0.46[a]	31
Pike	Downstream from the same plant	5.96[b]	31

[a] Average of 10 samples.
[b] Average of 17 samples.

mercury than those caught upstream (Table II). The chlor-alkali industry has been found to be the most important polluter of water in Canada.[36] Commercial fishing in many of these Canadian waters has been banned, as mercury concentrations in fish appeared to be dangerous to human health.

Other commercial uses of mercury such as in electrical equipment, paints, high-intensity lights, pharmaceutical preparations, etc., also release some mercury into the environment. However, this is a widely dispersed, low-level source which does not appear to be immediately hazardous.

IV. USE OF MERCURY IN AGRICULTURE

Compared to the tonnage of mercury used for industrial purposes the amount used in agriculture is rather small. Only 3.4% (204,000 lb) of all mercury consumed in the U.S.A. in 1969 was used in agriculture. In the same year the Canadian consumption of mercury was 300,000 lb of which 25,000 lb (8.3%) was used in agriculture.[37] On a worldwide basis about 2100 tons of mercury compounds were used in agriculture in 1965.[38] Considering the fact that the mercury content of mercurial compounds used in agriculture is usually less than 1% for seed dressings and 10 to 40% for orchard sprays, the absolute amount of mercury used in worldwide agriculture would be a small part of the 10,000 tons of mercury used per annum. Japan is the largest consumer of mercurial compounds in agriculture and in 1965 Japan used 1600 tons of the 2100 tons of these compounds used all over the world.[38]

Mercury compounds are used in agriculture to control fungus diseases of seeds, bulbs, plants, fruits and vegetation.[38] In Europe these compounds are used primarily to control seed-borne diseases while in North America they are used to control soil-borne fungi as well as seed-borne diseases. The rate of application of mercury to control seed-borne diseases is very small. Thus only about 0.5 g of mercury per acre is used as a seed dressing to control diseases in cereal crops in Canada.[39] Larger amounts (12 to 340 g Hg/acre) are used to control diseases in fruits and still more to control diseases in turf grass.[38]

V. MERCURY RESIDUES IN FOOD

All living things contain some traces of natural mercury but the use of mercury in agriculture has raised these levels in some cases. The extent to which this has happened is only discussed briefly here as the pertinent literature has already been reviewed by Smart.[38]

The quality of the discussion on any residue problem depends upon the accuracy of the analytical data on which it is based. Unfortunately it appears that there can be considerable variation in the apparent mercury content when determined by different laboratories using similar or different methods.

Table III shows the mercury contents of several samples of wheat and flour as determined by six laboratories.[40] The results obtained by Jervis et al.[41] are of special significance since these results received wide publicity in the mass media and were quoted as evidence in the U.S. Senate hearings of the Hart subcommittee.[42] After the report of Jervis et al. had been issued the same samples were analyzed again by four laboratories using the neutron activation method and one using wet oxidation and atomic absorption spectrometry. "In all some 51 analyses were carried out on these samples, 26 chemically and 25 by neutron activation", and "none of these values amounted to 10% of those reported by Jervis et al., most were less than 3%.[40] A further example of analytical variation is given in Table IV. Two laboratories, both using the neutron activation method, analyzed the same 13 samples of pheasant tissue. Laboratory A estimated that Specimen No. 15 contained 0.812 ppm of mercury, whereas the other laboratory gave a figure of 0.05 ppm for the same sample. This is an extreme example, but 2- to 4-fold differences are quite common.

This point is not raised to question the validity of all the residue data in the literature but to emphasize that there are limitations on analytical methods and human skills which must be taken into account when dealing with residue problems.

TABLE III
Comparative mercury analysis of wheat and flour (ref. 40)

Sample No.		Mercury (ppm)		
		Neutron activation		
		Jervis *et al.* (ref. 41)	Other laboratories	Atomic absorption
18464	Wheat	0.079	0.007[b]	0.005
20445	Wheat	0.30	0.014[b], 0.004[a]	0.007
18465	Wheat	0.40	0.012[b]	0.007
20389	Wheat	0.34	0.016[b], 0.010[a]	0.009
32573	Flour	0.38, 0.29	0.010[b], 0.015[a] 0.02[c], 0.011[d]	0.007
32574	Flour	0.26, 0.14	0.005[b], 0.013[a]	0.007
32575	Flour	0.22	0.008[b], 0.036[a]	0.005

[a] Atomic Energy of Canada Ltd.
[b] Oak Ridge National Laboratory.
[c] F.D.A. (U.S.), 9 replicates.
[d] Gulf General Atomics.

TABLE IV
Mercury content of pheasant tissues as found by two laboratories using the neutron activation analysis method (ref. 37)

Specimen No.	Mercury (ppm)	
	Laboratory A	Laboratory B
1	0.006	0.050
9	0.035	0.020
15	0.812	0.050
20	0.033	0.026
21	0.012	0.028
23	0.029	0.034
24	0.015	0.052
25	0.017	0.013
27	0.014	0.031
42	0.027	0.028
43	0.106	0.037
48	0.014	0.021
89	0.007	0.025

a. Cereal grains

It has been claimed by Löfroth and Duffy[6] that mercury will translocate from the treated seed to the harvested grain. Recently James et al.[43] have also claimed that wheat from plants grown from treated seed contained more mercury than wheat grown from untreated seed. These claims do not seem to be supported by the experimental evidence. The data of James et al., when analyzed by the authors, were not statistically significant at the 5% level but the authors still claimed that the differences in mercury content between grain from treated and untreated seed were real.

According to two unpublished sources cited by Smart[38] the mercury content of harvested wheat or barley does not seem to be affected by mercury seed dressings. Saha et al.[39] also found that wheat and barley contained from 0.008 to 0.016 ppm of mercury whether it was grown from treated or untreated seed.

The background level of mercury in rice (0.2 ppm) is much higher than that in wheat or barley (0.01 to 0.02 ppm).[38] This background level may be further increased by the use of mercurial compounds during cultivation. Tomizawa[44] found 0.1 to 1.0 ppm of mercury in polished rice from paddies where phenylmercury acetate had been applied, compared to 0.23 to 0.24 ppm in rice from untreated fields.

b. Fruit

Mercury background levels in apples and pears are normally 0.04 ppm or less.[38] This level is increased to about 0.1 ppm when these crops are treated with mercury compounds used in accordance with good agricultural practice.

c. Tomatoes

Tomatoes from untreated plants have been found to contain up to 0.02 ppm of mercury from natural sources.[38] Multiple applications of phenylmercury chloride or phenylmercury salicylate on tomatoes in glasshouses can result in increasing the residues up to 0.1 ppm.[45] Smart found a maximum residue of 0.11 ppm mercury in 13 samples of commercially-grown tomatoes which had received up to eight treatments.

d. Potatoes

Background levels of mercury in potatoes are about 0.01 ppm.[38] The application of four sprays of phenylmercury chloride-copper oxychloride at the normal rate in August produced residues of 0.01 and 0.05 ppm respectively in the whole tuber.[46,47] More than 60% of the residue was in the peel.

e. Meat and eggs

The mercury content of pork, veal, ox, and reindeer meat was investigated by Westöö.[48] The mean residue levels were: Swedish pork cutlets, 0.030 ppm; Danish pork cutlets, 0.003 ppm; Swedish bacon, 0.018 ppm; Danish bacon, 0.004 ppm; Swedish pigs' liver, 0.060 ppm; Danish pigs' liver, 0.009; Swedish ox meat, 0.012 ppm; reindeer meat, 0.013 ppm. It appears that Swedish products contained more mercury than those from Denmark, indicating the presence of more mercury in animal feed in Sweden. The mercury content of eggs sold in Sweden was also reported to be higher than other European countries and it was implied that mercury used as seed dressings in Sweden led to the contamination of hens' eggs.[32] Later experiments showed that the mercury content of eggs of hens fed grain grown from untreated seed was about 0.01 ppm and that similar levels were present in eggs of hens that were fed grain grown from methoxyethylmercury-treated seed. However, an average of 0.027 ppm mercury was present in the eggs of hens that were fed grain grown from alkylmercury-treated seed. Since the feed did not consist entirely of grain and contained significant amounts of fish meal whose mercury content was not known, the significance of this study cannot be ascertained. A more plausible explanation for the higher levels of mercury in the eggs in Sweden may be the widespread contamination of fish in that country. Since fish protein is a common constituent of chicken diets, elevated levels of mercury could be expected in the feed and subsequently in the eggs. Whatever the reason for higher mercury levels in Swedish meat and eggs, they are below 0.05 ppm and do not appear to be hazardous to human health.

f. Fish

Residues of mercury in fish probably constitute the most serious of the mercury pollution problems facing many countries. However, although this may increase the mercury content of people's food, the source of contamination is industrial rather than agricultural. Numerous reports have been published in the literature on the mercury content of fish from waters with known sources of pollution. These reports have been reviewed by Saha.[13] In general, the background levels for mercury in fish are 0.2 ppm or less[38] but there have been numerous reports where mercury levels in fish from apparently unpolluted waters exceeded 0.5 ppm.[13]

g. Whole diet

It is apparent from the above that the use of mercury in agriculture may lead to some increase in the mercury content of some foodstuffs. However, the average diet consists of a mixture of foodstuffs and the important question is whether the use of mercury in agriculture has led to any appreciable increase in the mercury content of the whole diet. Stock and Cucuel[21] reported the mercury content of some foods in 1934, and subsequent studies were made by Gibbs et al.[49] in 1941 and Goldwater[50] in 1964. The comparison of these figures (Table V) suggests that in general the mercury content of foodstuffs has not changed significantly over the past 30 years, although the agricultural use of mercury increased considerably during this period. These results may simply indicate that only a small fraction of the food supply is exposed to mercury compounds during production.

Westöö[51] reported the total mercury content of 14 daily diets, including certain beverages, from Stockholm, Sweden. Mercury levels in 12 diets without fish were between 0.004 and 0.013 ppm. This is equivalent to a

TABLE V

Mercury in food (ppm)

	Stock and Cucuel (ref. 21) Germany	Gibbs et al. (ref. 49) U.S.A.	Goldwater (ref. 50) U.S.A.
Meats	0.001–0.067	0.0008–0.044	0.001–0.15
Fish	0.02–0.18	0.0016–0.014	0–0.06
Fresh vegetables	0.002–0.004	0	0–0.06
Fresh milk	0.0006–0.004	0.003–0.007	0.008
Grains	0.02–0.036	0.002–0.006	0.002–0.025
Fresh fruit	0.004–0.01	—	0.004–0.03
Egg	0.002	0	—
Beer	0.00007–0.0014	—	0.004

mercury intake of 4–20 mcg/day. In two diets containing 16 g and 165 g of fish the mercury levels were 0.003 and 0.014 ppm, corresponding to a daily intake of 4 and 33 mcg, respectively.

Abott and Tatton[52] found no detectable mercury residues in composite food samples from England and Wales. They estimated from the detection limit of their analytical method that the daily intake would be less than 14 mcg/day.

Sommers[53] discussed the daily intake of traces of heavy metals by Canadians. Composites of the twelve major food classes, prepared as by the average housewife, were analyzed by atomic absorption spectrophotometry and from this the average daily intake was calculated. Compared to the average daily intakes for other heavy metals, for example, zinc at 19,900 mcg/person, copper at 2200, lead at 139, the daily intake of mercury was very low. In fact, the daily intake of mercury in the whole diet, 20 mcg/person, was so low that it had to be calculated as the maximum that could occur if all the foodstuffs had been at or just below the detection limit of 0.02 ppm. In addition, the Canadian Food and Drug Directorate analyzed 1400 food samples, excluding fish, for mercury, and only isolated samples were found to be higher than 0.1 ppm.

On the basis of the data available it does not appear that the mercury content of the average diet has been appreciably changed by the agricultural use of mercury. In fact, the total mercury from all sources, natural or man-made, appearing in the average diet seems to be extremely low. Obviously special cases may exist where there is a real hazard. For example, if a large part of a person's diet consisted of fish taken from mercury-contaminated waters he would be exposed to a high risk as in Minamata and Niigata but, in general, the risk would appear to be small.

VI. INCIDENTAL HAZARDS FROM THE USE OF MERCURY IN AGRICULTURE

Although the proper use of mercury in agriculture may not be directly hazardous to man, there have been a number of cases of poisoning due to misuse and some unfortunate ecological effects. These have largely arisen through the use of mercurial seed dressings, particularly with methylmercury compounds, on grain.

Poisoning from occupational exposure to methylmercury compounds has been reported for laboratory workers,[54-56] workers in factories producing seed-dressing compounds,[55-61] farmers who dressed seed for sowing,[62-64] and also workers in pulp mills.[64,65] Inhalation of vapour or dust was the main cause of exposure but contamination of clothes and skin with liquid seed-dressing compounds may have occurred also. In all, 20 cases of poisoning were reported in these studies, four of them fatal. There was another fatal poisoning where the person was treating wood with methylmercury.[55,56]

Poisoning from nonoccupational exposure to methylmercury seed dressings has also been reported. A few cases occurred in Sweden during the 1940s when some people ate treated seed. Engelson and Herner[66] reported a case of poisoning in a 13-month-old boy who was fed porridge made of flour from treated seed, probably daily, for a period of four months. They also reported a case of suspected prenatal poisoning in a child whose mother had eaten dressed grain during her pregnancy.

Ordonez et al.[67] reported mercury poisoning in 45 children and adults in

Guatemala who had consumed treated seed containing 17 ppm of mercury. Twenty persons died from this accident. As stated before, similar accidents occurred in Iraq and Pakistan, where people ate dressed seed.[2] Storrs *et al.*[11,12] reported the poisoning of two children and a young woman from New Mexico, U.S.A., after eating meat from a pig that was fed seed treated with methylmercury dicyandiamide. The treated seed had 33 ppm of mercury and the pork had 28 ppm. Although seven of the nine members of the family ate the same meat for a period of four months, only three persons showed definite signs of mercury poisoning. These incidents arose because of improper use and this suggests that where there is a choice of compounds the chemical least prone to misuse should be recommended.

The ecological dangers of the agricultural use of mercury are more serious. The first suggestion that the "proper" use of mercury in agriculture might be dangerous came from Sweden. In 1960 attention was drawn to the possibility that game birds and other wildlife were being poisoned by mercurial seed dressings. Borg *et al.*[68] found very high levels of mercury (up to 270 ppm) in the livers and kidneys of birds found dead in the Swedish countryside. The mercury in these birds was attributed to the mercurial seed dressings used in Swedish agriculture.[69] It was shown that some treated seed (about 1.7 kg/ha) was left on the surface of the fields at planting and was thus available to seed-eating birds. Berg *et al.*[70] also determined the mercury content of feathers from several species of Swedish birds collected by Swedish museums between 1840 and the present time. It was shown that mercury levels remained fairly constant within species during the period 1840 to 1940. But these levels increased sharply, 10-to 20-fold, after about 1945 when alkylmercury seed dressings came into use. It was also observed that the use of inorganic mercury in agriculture before 1930 and phenylmercury from 1930 to 1940 had not increased the mercury content of these birds. After 1966 Sweden replaced the alkylmercury seed dressings with alkoxyalkylmercury compounds and there was a noticeable decline in mercury residues in the seed-eating birds and their predators.[33]

Recent studies in Canada have also shown the contamination of seed-eating birds and their predators with mercury.[37,71] Mercurial seed dressings used in agriculture have been considered as the major cause of this contamination. Average mercury levels in seed-eating rodents, songbirds, and upland game birds collected from areas where mercurial seed treatments had been used were 1.25, 1.63, and 1.88 ppm, respectively. The corresponding levels in similar specimens collected from an untreated area were significantly lower: 0.18, 0.03, and 0.35 ppm, respectively. There were also higher levels of mercury in tissues of predatory birds and their prey in Alberta as compared to those from Saskatchewan, reflecting the use of much less treated seed in Saskatchewan.

On the other hand, low levels of mercury (0.007 to 0.075 ppm) were found in the tissues of strictly herbivorous animals such as cow, horse, deer, moose, etc.[33] These studies indicate that, although the use of mercury seed dressings, especially alkylmercury compounds, can cause widespread contamination of seed-eating birds and their predators, such uses may not lead to any significant increase in the mercury content of other terrestrial fauna. Moreover, replacement of alkylmercury compounds with alkoxyalkylmercury compounds for seed treatment reduces the contamination of seed-eating birds.[33]

There is no evidence that alkylmercury seed dressings make any significant contribution to the pollution of soil and water. It has been stated before that only about 0.5 g/acre is used as a seed dressing. This amount (about 0.0005 ppm based on the weight of an acre of soil 6 in deep) is insignificant when compared to the natural mercury content of soil and less than the amount of mercury received by soil through precipitation.[27] Recently a study was carried out by Sand et al.[28] on the effect of the use of mercury seed dressings on the residue level in soil. Half the soil samples were collected from areas where mercury-treated seed had been used and the other half were from areas where it had not been used. A significant difference was found in mean levels between areas of use and non-use, with the higher residues occurring in areas where mercury had not been used. These results indicate that the amounts of mercury added to soils by seed dressings may be less than the variation in background levels occurring naturally.

VII. TOXICITY OF MERCURY RESIDUES

Residue data, in themselves, are of little value unless one can make some reasonable assessment of the toxicological hazards involved. The acute toxicity of a compound, where one or more massive doses produce an immediate effect, can be measured fairly readily. Chronic toxicity, the effect of many small doses over a long period, is a more subtle danger and very difficult to assess. In addition, one is concerned not only with the effect on the adult but also with possible effects on the reproductive cells, the embryos and the young. The complications are such that we can probably never say that a compound is absolutely and completely safe under all conditions; at the very best we can only say that, within the limits of present knowledge, the risk appears to be small.

The toxicity of mercury residues, from any source, will depend on the chemical nature of the mercury compound. Metallic mercury when ingested orally is not very toxic but chronic exposure to its vapour can be hazardous.[72] The soluble inorganic salts of mercury have been known to be toxic for a long time. Organomercury compounds are more toxic to man than the inorganic salts of mercury. However, organomercurials have different degrees of toxicity. Alkoxyalkylmercury compounds have low toxicities, whereas alkylmercury, particularly methylmercury compounds, are highly toxic.[2,73]

The real danger to living organisms comes from the presence of alkyl-mercury compounds, principally methylmercury, in the environment. The mercury-containing effluents of chlor-alkali plants, pulp and paper mills, pastics industries, etc., are largely responsible for high local concentrations of mercury in lakes or rivers. Fish, plants or animals cannot, themselves, convert other mercury compounds into methylmercury,[13] but micro-organisms in bottom muds can do so.[33,76] This methylmercury is then taken up by aquatic organisms and concentrated in the food chains. The result is high concentrations of mercury in fish from contaminated waters, mainly in the form of methylmercury.[32,74,75]

Methyl and other alkylmercury compounds are used in agriculture as seed dressings. Accidental poisoning from this use can occur through careless handling of the chemicals or from eating treated grain. In addition, treated seed is hazardous to seed-eating birds and their predators. Since these com-pounds are not translocated into the harvested grain from treated seed, there is little or no possibility of contamination of man's food.

Poisoning from methylmercury is characterized by sensory disorders, ataxia, concentric constriction of the visual fields, impairment of hearing, symptoms from the autonomic nervous system and extrapyramidal system as well as "mental disturbances".[8,9,11,12,55,56,62] The illness is often called the Minamata disease.

Methylmercury compounds are almost completely absorbed from the gastro-intestinal tract and they have considerable stability in the animal body.[17,73] The biological half-life of these compounds in human beings is about 70 days. They accumulate in the central nervous system and concentrations of 8 ppm or over in the brain can cause poisoning. Methylmercury compounds can pass freely through the placenta and damage the central nervous system of the foetus. They have also been suspected of causing certain genetic effects. Finally, there appears to be no satisfactory therapy for methylmercury poisoning.

In view of these factors, concern about methylmercury compounds in the environment is quite justified and tolerance levels should be established for methylmercury in food and particularly in fish, as this is the principal source of methylmercury in our food. Two estimates of "allowable daily intake" (ADI) of methylmercury have been proposed by Swedish scientists.[77] One ADI, equivalent to 0.7 mg mercury (from methylmercury) per week, was calculated from the relationship between intake and the mercury level in the blood of high consumers of fish in Sweden. The second ADI, equivalent to 0.42 mg Hg/week, equals the estimated equilibrium in a subject with a body burden of 6 mg Hg due to methylmercury (one-tenth of the estimated toxic or near-toxic level). From these considerations a "practical residue limit" of 1 ppm mercury (as methylmercury) has been allowed in Sweden. Japan also allows 1 ppm Hg in fish. The *per capita* consumption of fish in Canada and the U.S.A. is 17 g/day as compared to 56 and 84 g/day in Sweden and Japan, respectively. Since the tolerance level of mercury in Canada and the U.S.A. is 0.5 ppm in fish, it is unlikely that even the heavy fish eaters in these two countries will ever receive enough methylmercury from fish to exceed the lower of the two proposed ADIs.

Apart from fish, man may still be exposed to traces of mercury in other food items. In 1966, two FAO and WHO expert committees proposed a "practical residue limit" of 0.05 ppm in food.[78] This level did not take into account the chemical nature or the background levels of mercury in different foods. This level is practical for most vegetables, wheat, barley, meat and eggs but impractical for rice, fish and possibly some fruits, as these latter commodities have background levels of mercury higher than this tolerance level. Although we know that most of the mercury in fish and animal products is present as methylmercury, we know little about the chemical nature of mercury compounds in other foods. Although organomercury compounds are used for the protection of crops such as apples, rice, tomatoes or potatoes, the terminal nature of the residues in these crops is not known. If the terminal residue is not very toxic (such as inorganic mercury), then more of it could be tolerated than when it is present in highly toxic form (i.e. methylmercury). Most mercury residue data do not differentiate between organic and inorganic mercury and give only the total mercury present in the sample. Research should be carried out to identify the terminal mercury compounds in food and also to develop analytical methods for determining such residues. Unless this is done it will be difficult to establish tolerances that reflect the toxicity of the residues. Meanwhile, the guidelines for allowable levels of mercury in food should be based on the background levels of mercury in a given commodity and also the levels occurring in that commodity from good agricultural practice.

The official guidelines for mercury residues in foodstuffs in many countries appear to be utterly unrealistic, as they do not even take into account the background levels. Thus, Australia has an official tolerance of 0.03 ppm of mercury in fruits and vegetables.[38] But since apples, for example, have a background level of 0.04 ppm of Hg, one might ask—are any apples sold there? The sale of food containing residues of mercury whose origin is mercury-containing pesticides is illegal in Germany.[38] One might suggest that it may be difficult, if not impossible, to distinguish the mercury from natural sources from that of pesticidal origin! Again there are countries who do not allow any mercury residues in food[38]—an utterly ridiculous and impractical proposition—as all food contains determinable quantities of mercury. Clearly, regulatory authorities have to give serious considerations to this matter and set tolerance levels that are practical to enforce and safe for human health.

VIII. SUMMARY

Most of the mercury in the environment is of natural origin. The surface of the earth receives about 100,000 tons of mercury annually through precipitation as compared to world production of 10,000 tons per year. The natural cycle of circulation of mercury on earth disperses it widely through rocks and soil, water, air, and the biosphere.

It has long been known that traces of mercury occur everywhere. The average mercury content of the earth's crust is about 0.5 ppm. The background level of mercury in soil is 0.1 to 2.0 ppm, in river water about 0.1 ppb, and sea-water up to 0.3 ppb.

All food contains traces of natural mercury. In fruits, the background levels are normally 0.04 ppm of mercury or less; in vegetables 0.02 ppm or less; in wheat and barley up to 0.02 ppm; in meat up to 0.05 ppm; and in rice about 0.2 ppm.

Only a small portion of the mercury used by man is utilized in agriculture. This use of mercury may increase the mercury content of some crops (such as apples, tomatoes, and potatoes) but in other cases it has no effect. Published data indicate that during the past 30 years there has been no significant increase in the over-all mercury content of foodstuffs sold in the market place. Studies on mercury content of total diets in many countries indicate that our daily intake of mercury through food is less than 20 mcg per person.

Soil receives less than 1 g of mercury/acre from seed dressings. This amount is insignificant when compared with the natural mercury content of soil and less than it receives through precipitation annually. Such uses also do not contribute to the pollution of water and fish with mercury.

Although the use of mercury in agriculture has not significantly affected the quality of agricultural products, soil and water, it has affected the seed-eating birds and their predators. Dressed seed left uncovered in the field or elsewhere may be eaten by seed-eating birds, resulting in very high levels of mercury in them and their predators.

The largest consumer of mercury is the chlor-alkali industry which is also the most serious polluter of the aquatic environment.

Alkylmercury, particularly methylmercury compounds, is highly toxic. Almost all the mercury in fish and animal products is present as methylmercury. Consumption of fish containing high levels of methylmercury can have serious consequences, as demonstrated in the Minamata and Niigata disasters in Japan.

The regulation of mercury levels in food is still a problem area. Firstly, where tolerance levels are set, they do not distinguish between the various compounds of mercury although these differ widely in toxicity. Secondly, the tolerances are sometimes unrealistic in that they can be below the background levels found in untreated plants or animals.

IX. Bibliography

1. G. T. Engel, Mercury. In *Kirk–Othmer Encyclopedia of Chemical Technology*, Vol. 13 (John Wiley & Son, New York, 1967), 2nd ed.
2. L. J. Goldwater, *Sci. Amer.* **224**, 15 (1971).
3. Anonymous, Environ. *Sci. Technol.* **4**, 890 (1970).
4. N. Grant, *Environment* **11**, 19 (1969).
5. L. Kurland, *World Neurology* **1**, 370 (1960).
6. G. Löfroth and M. E. Duffy, *Environment* **11**, 10 (1969).
7. N. Fimreite, *Environ. Pollut.* **1**, 119 (1970).
8. Minamata Report. M. Kutsuna (Ed.), *Minamata disease*. Study Group of Minamata Disease (Kumamoto University, Japan, 1968).
9. Niigata Report. *Report on the cases of mercury poisoning in Niigata* (Ministry of Health and Welfare, Tokyo, 1967).
10. K. Irukayama, T. Kondo, F. Kai, and M. Fujiki, *Kumamoto Med. J.* **15**, 57 (1962).
11. B. Storrs, J. Thompson, G. Fair, M. S. Dickerson, L. Nickey, W. Barthel, and J. E. Spaulding, *Morbidity and Mortality* **19**, 25 (1970).
12. B. Storrs, J. Thompson, L. Nickey, W. Barthel, and J. E. Spaulding, *Morbidity and Mortality* **19**, 169 (1970).
13. J. G. Saha, *Residue Rev.* **42**, 1972.
14. R. Miskus, DDT. In *Analytical Methods for Pesticides, Plant Growth Regulators, and Food Additives*, edited by G. Zweig (Academic Press, New York, 1964).
15. J. W. Mellor, *A Comprehensive Treatise on Inorganic and Theoretical Chemistry* (Longmans, Green and Co. Ltd., London, 1929).
16. N. A. Ozerova, *Primary Dispersion Halos of Mercury* (Academy of Sciences USSR, Moscow, 1962).
17. A. A. Saukov, *Geokhimiya rtuti* (Geochemistry of mercury). Tr. Inst. Geol. Nauk. Akad. Nauk SSSR, 78, Min-Geokhim. Ser. No. 17, 1 (1946).
18. J. M. West, Mercury. In *Minerals Year Book* 1968, Vol. I–II (U.S. Govt. Printing Office, 1969).
19. A. A. Saukow, *Geo. hemie*. (Verlag Technik, Berlin, 1953).
20. J. R. Partington, *History of Chemistry*, Vol. 3 (McMillan & Co. Ltd., London, 1962), pp. 77 and 640.
21. A. Stock and F. Cucuel, *Naturwissenschaften* **22**, 390 (1934).
22. F. Heide, H. Lerz, and G. Bohm, *Naturwissenschaften* **44**, 441 (1957).
23. K. Hosohara, *Nippon Kagaku Zasshi* **82**, 1107 (1961).
24. R. C. Weast, *Handbook of Chemistry and Physics* (The Chemical Rubber Co., Cleveland, Ohio, 1968, 49th ed.
25. K. Rankama and Th. G. Sahama, *Geochemistry* (The Univ. of Chicago Press, Chicago, 1950).
26. J. T. Martin, *Analyst* **88**, 413 (1963).

27. A. Anderson, *Grundforbattring* **20,** 95 (1967).

28. P. F. Sand, G. B. Wiersma, H. Tai, and L. J. Stevens, *Pestic. Monitor. J.* **5,** 32 (1971).

29. H. V. Warren, R. E. Delavault, and J. Barakso, *Econ. Geol.* **61,** 1010 (1966).

30. L. Hannerz, Inst. of Fresh Water Research, Drottinghlom, Sweden, Report No. 48 (1968).

31. E. G. Bligh, *Mercury and the Contamination of Freshwater Fish.* Fish. Res. Bd. Can., Manuscript Rep. Ser. No. 1088, Winnipeg, Man., April (1970).

32. G. Westöö, Methylmercury compounds in animal foods. In *Chemical Fallout*, edited by M. W. Miller and G. G. Berg (Charles C. Thomas Publ., Springfield, Ill., 1969), p. 75.

33. A. G. Johnels and T. Westermark, Mercury contamination of the environment in Sweden. In *Chemical Fallout*, edited by M. W. Miller and G. G. Berg (Charles C. Thomas Publ., Springfield, Ill., 1969), p. 221.

34. G. Wobeser, N. O. Nielsen, R. H. Dunlop, and F. M. Atton, *J. Fish. Res. Bd. Can.* **27,** 830 (1970).

35. A. K. Sumner and J. G. Saha, Unpublished results.

36. E. G. Bligh, *Proc. of the Royal Soc. of Canada Internat. Symp. on Mercury in Man's Environment*, Ottawa (1971).

37. J. B. Gurba, *Proc. 18th Annual Meeting and Conf. of the Canadian Agricultural Assoc.*, Jasper, Alta (1970).

38. N. A. Smart, *Residue Rev.* **23,** 1 (1968).

39. J. G. Saha, Y. W. Lee, R. D. Tinline, S. H. F. Chinn, and H. M. Austenson, *Can. J. Plant Sci.* **50,** 597 (1970).

40. E. Sommers, *Proc. of the Royal Soc. of Canada Internat. Symp. on Mercury in Man's Environment*, Ottawa (1971).

41. R. E. Jervis, D. Debrum, W. LePage, and B. Tiefenbach, *Mercury residues in Canadian foods, fish, wildlife.* National Health Grant Project No. 605-7-510 (Univ. of Toronto, 1970).

42. *Pesticide Amendments to Hazardous Substances Act.* Hearings before Senate Subcommittee on S. 3866, Serial 91–79 (U.S. Govt. Printing Office, 1970), p. 59.

43. P. E. James, J. V. Lagerwerf, and R. F. Dudley, *Proc. of the Internat. Symp. on Identification and Measurement of Environmental Pollutants*, Ottawa (1971).

44. C. Tomizawa, *Shokuhin Eiseigaku Zasshi* 7, 26 (1966).

45. H. M. Stone and P. J. Clark, *N.Z. J. Sci.* **1,** 373 (1958).

46. R. G. Ross and D. K. R. Stewart, *Can. J. Plant. Sci.* **44,** 123 (1964).

47. D. K. R. Stewart and R. G. Ross, *Can. J. Plant Sci.* **42,** 370 (1962).

48. G. Westöö, *Var Föda* **18,** 85 (1966).

49. O. S. Gibbs, H. Pond, and G. A. Hansman, *J. Pharmacol.* **72,** 16 (1941).

50. L. J. Goldwater, *J. Roy. Inst. Public Health* **27,** 279 (1964).

51. G. Westöö, *Var Föda* **4,** 1 (1965).

52. D. C. Abott and J. O'G. Tatton, *Pestic. Sci.* **1,** 9 (1970).

53. E. Sommers, *Proc. of the Internat. Symp. on Identification and Measurement of Environmental Pollutants*, Ottawa (1971).

54. E. Franke and K. D. Lundgren, *Arch. Gewerbepath. Gewerbehyg.* **15,** 186 (1956).

55. K. D. Lundgren and A. Swensson, *Nord. Hyg. T.* **29,** 1 (1948).

56. K. D. Lundgren and A. Swensson, *J. Industr. Hyg.* **31,** 190 (1949).

57. A, Ahlmark, *Brit. J. Industr. Med.* **5,** 117 (1948).

58. T. Herner, *Nord. Med.* **26,** 833 (1945).

59. D. Hunter, R. R. Bomford and, D. S. Russell, *Quart. J. Med.* **33,** 193 (1940).

60. J. J. G. Prick, A. E. H. Sonnen, and J. L. Slooff, *Wetenschappen* **70,** 150 (1967).

61. J. J. G. Prick, A. E. H. Sonnen, and J. L. Slooff, *Wetenschappen* **70,** 170 (1967).

62. G. Ahlborg and A. Ahlmark, *Nord. Med.* **41,** 503 (1949).

63. G. Bloom, K. D. Lundgren, and A. Swensson, *Nord. Hyg. T.* **36,** 110 (1955).

64. K. D. Lundgren and A. Swensson, *Amer. Industr. Hyg. Assoc. J.* **21,** 308 (1960).

65. S. Freyschuss, O. Lindstrom, K. D. Lundgren, and A. Swensson, *Svensk Papperstindning* **61,** 568 (1958).

66. G. Engelson and T. Herner, *Acta Paediat. Scand.* **41,** 289 (1952).

67. J. V. Ordonez, J. A. Carrilo, M. Miranda, and J. L. Gale, *Boletin de la Oficina Sanitaria Panamericana* **60,** 510 (1966).

68. K. Borg, H. Wanntorp, K. Erne, and E. Hanko, *J. Appl. Ecol. Suppl.* **3,** 171 (1966).

69. S. Tejning, *Oikos* **18,** 334 (1967).

70. W. Berg, A. G. Johnels, B. Sjostrand, and T. Westermark, *Oikos* **17,** 71 (1966).

71. N. Fimreite, R. W. Fyfe, and J. A. Keith, *Can. Field-Naturalist* **84,** 269 (1970).

72. P. L. Bidstrup, *Toxicity of Mercury Compounds* (Elsevier Publishing Co., Amsterdam (1964).

73. Anonymous, *Arch. Environ. Health* **19,** 891 (1969).

74. A. K. Sumner and J. G. Saha, *Environ. Lett.* **2,** 167 (1971).

75. C. A. Bache, W. H. Gutenman, and D. J. Lisk, *Science* **172,** 951 (1971).

76. S. Jensen and A. Jernelov, *Nature* **223,** 753 (1969).

77. F. Berglund and M. Berlin, Risk of methylmercury cumulation in man and mammals and the relation between body burden of methylmercury and toxic effects. In *Chemical Fallout*, edited by M. W. Miller and G. G. Berg (Charles C. Thomas Publ., Springfield, Ill., 1969).

78. *FAO Working Party on Pesticide Residues and WHO Expert Committee on Pesticide Residues.* WHO Techn. Rep. Ser. No. 370, FAO Agricultural Studies No. 73, FAO, Rome (1967).

The Microdetermination of Mercury and Organomercury Compounds in Environmental Materials

J. F. UTHE and F. A. J. ARMSTRONG

Department of the Environment, Fisheries Research Board of Canada, Freshwater Institute, Winnipeg, Manitoba R3T 2N6, Canada

Contents

1. INTRODUCTION

In late 1969, results of investigations[1] carried out at the University of Saskatchewan, Canada, were communicated to officials of the Federal Department of Fisheries and Forestry. Resultant action, involving the closure of many lakes to fishing, and seizure and destruction of large quantities of fish in which the average total mercury content exceeded 0.5 mcg/g wet weight, finally managed to bring the necessary public awareness of mercury pollution on this continent.[2]

21

This was by no means the first discovery of mercury pollution in the aquatic environment. In 1958 a number of cases of a mysterious neurological disease were reported in the Minamata area of Japan.[3] Out of a total of 121 clinical cases there were 46 deaths.[4] A disproportionately high number of infant cases[5] were seen and among the survivors severe permanent disability was the rule.[6,7] Although the disease was fairly rapidly correlated to the eating of large quantities of fish and shellfish from Minamata Bay,[7] and fishing stopped, it was not until 1964 that the etiology and the causative agent of the disease were established. Minamata disease was shown to be caused by the action of the methylmercury radical (CH_3Hg^+) on the central nervous system.[8,9] Levels as high as 85 mcg/g dry weight have been reported in shellfish from Minamata bay.[10] In 1965 another outbreak of Minamata disease occurred in Niigata, Japan. This time there were 47 cases and 6 deaths.[11] In both of these cases the source of the methylmercury was thought to be discharge of methylmercury contaminated industrial wastes into the waters of these two areas.

In the early 1960s Sweden became aware of a mercury problem in its environment through the finding of high levels of mercury in the bodies of seed-eating birds.[12,13] The source of this mercury was thought to be the ingestion of seeds which had been dressed with mercurials prior to sowing. In 1964 high mercury levels were found in the flesh of northern pike (*Esox lucius*) from freshwater areas of Sweden.[14]

In 1966 Westöö[15] showed the presence of methylmercury in Swedish fish. In fact, the major portion of the mercury in these fish was present as methylmercury.[16,17] What was probably the most shocking discovery of this period was that of Jensen and Jernelov[18] who showed the widespread occurrence of micro-organisms capable of transforming inorganic and other organic mercurials, such as phenylmercuric acetate, into methylmercury. The toxicological dangers of the discharge of mercury into the environment became obvious.

The rapid elucidation of the mercury problem in North America has led to the development and modification of a variety of methods for the quantitative determination of environmental mercury levels.[19-22] Although the Association of Official Analytical Chemists (AOAC) had recommended a spectrophotometric method[23] based on the determination of the mercury-dithizone complex formed after release of the mercury from the tissue by wet digestion, the method was not readily adaptable to the determination of mercury content in large numbers of samples and lacked sufficient sensitivity to enable use of small amounts of material ($\lesssim 1$ g). Methods for the determination of methylmercury are much fewer in number and are generally based on the original finding of Gage[24] that methylmercury bound in tissues could be rendered soluble in organic solvents by the action of strong halide acids. Some work also has been done on methods which measure the inorganic and organic mercury content of a material.

In this review the current methods used for analysis of mercury in environmental samples will be reviewed. In the main this review will centre around the use of atomic absorption spectrophotometry (AAS) for the determination of total mercury content and use of gas chromatographic methods for the determination of methylmercury. No attempt will be made to describe in exact detail methods covered in fairly recent reviews.[4,25-27]

2. DETERMINATION OF TOTAL MERCURY

a. Digestion procedures

Destruction of organic matter is necessary in chemical methods of analysis where mercury (II) ion is required, or in neutron-activation procedures when activated mercury has to be separated from other radionuclides. Mercury is completely volatilised by dry ashing;[28] this can be exploited in some atomic absorption methods.

Wet-ashing procedures of varying severity have been recommended. The most rapid and rigorous ones use perchloric acid. For safety an excess of nitric acid must be used with it at the start of the oxidation.[29] Gorsuch[28] showed that little mercury was lost when heated alone with mixtures of nitric, sulphuric, and perchloric acids, but that losses increased when organic matter was present, being greatest with nitrogen-containing compounds. When vapours were condensed and collected, recoveries of 95% or more were obtained. The S.A.C. Committee[30] did not recommend use of perchloric acid; in their procedure, and in that of the AOAC[31] in which only nitric and sulphuric acids are used, some fat may remain undigested. It may be removed by filtration with negligible loss of mercury.[30]

Many methods use nitric and sulphuric acids under reflux and complete the oxidation with permanganate. The authors have usually been able to demonstrate satisfactory recoveries of mercury. Jacobs et al.[32] and Uthe et al.[19] have shown satisfactory recovery of mercury at 50–60°C with sulphuric acid and permanganate, and this has been confirmed by Thorpe.[21] This low-temperature digestion is made more rapid if nitric acid is added, and if the permanganate is allowed to react at higher temperature but below the boiling point.[33]

Other oxidants used have been: H_2SO_4 with H_2O_2 for paints;[34] H_2SO_4 with $KMnO_4$ for urine;[35,36] H_2SO_4 with HNO_3 and H_2O_2 for plant material;[37] H_2SO_4, $HClO_4$, and H_2O_2 for vegetation;[38] HNO_3 and $HClO_4$ for rice;[39] HNO_3, $HClO_4$, and Se for potatoes[40] and apple peel;[41] HNO_3 and V_2O_5 for biological materials;[42] and H_2SO_4, HNO_3, and Se for apples.[43]

Samples of relatively low organic content, such as urine or water, are often given less drastic oxidation. HNO_3 or H_2SO_4 with $KMnO_4$ is frequently used.[35,44–47] Reith and van Dijk[48] digested urine and tissue samples with HCl and $KClO_3$, with Mn as catalyst. Liebmann and Hempel[49] used sulphuric acid alone. Becknell et al.[50] oxidised water samples with chlorine only, since this is reported to break alkyl-mercury bonds.[51]

The oxygen flask combustion method, with acid or acid permanganate to absorb mercury vapour, has been used successfully.[52,53] Incombustible materials such as sediments can be handled if cellulose is added.[54]

Lidums and Ulfvarson[55] dispersed heterogenous organic materials and prepared average solutions for analysis by heating with sodium hydroxide.

b. Dithizone

Colorimetric determination with dithizone is most useful where large numbers of analyses are not required. No special equipment other than a spectrophotometer is needed. To attain precision better than 0.1 ppm, sample weights of 5 g and upward are required and the digestion process becomes tedious.

Dithizone (diphenyl thiocarbazone) can react with mercury to form four coloured complexes. For analytical use the primary compound with mercury (II)

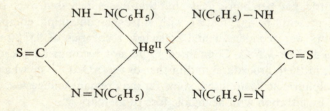

is usually formed by shaking the test solution with a small excess of a carbon tetrachloride or chloroform solution of the reagent. In these solvents the solution is orange-yellow with maximum absorbance at 490 nm and $\varepsilon =$ approx. 65,000.[56] Some organomercurial compounds can also form dithizonates; for example, methylmercuric dithizonate has maximum absorbance at 476 nm in chloroform with $\varepsilon = 32,000$.[57] Kurayaki and Kusumoto[58] describe separation of inorganic, alkyl and aryl mercury compounds by chromatography on ion exchange paper, with detection of spots by means of dithizone. Dithizone itself in chloroform or carbon tetrachloride gives green solutions with maximum absorption at 620 nm.

Primary mercuric dithizonate is almost unique in being formed rapidly and quantitatively at high acidity. Koroleff[59] found that it was formed in 12N HCl solution, but it is more usual to employ solutions 0.1 to 1.0N. Koroleff also found (contrary to the statement by Welcher[60]) that chloride inhibited the extraction of the dithizonate at high acidity. The ability of mercury to react in strong acid allows separation from most other metals, and only silver and copper are likely to interfere. Extraction in the presence of chloride prevents extraction of silver, or the extract containing silver may be shaken with chloride or thiocyanate to decompose silver dithizonate. Copper will normally be extracted with mercury, but in chloroform solution the copper dithizonate is formed only slowly. The use of this solvent minimised copper interference, which can be further suppressed by cobalticyanide.[34] However, at high copper concentrations cobalticyanide produces a precipitate.

If the extract containing copper and mercury dithizonates is extracted with acid thiosulphate, mercury is transferred to the aqueous phase, and the organic layer containing copper can be rejected.[61] Excess thiosulphate may then be decomposed with hypochlorite and the mercury re-extracted with dithizone.

Iodide may be used to complex copper, if the acidity is carefully controlled, so that only mercury is extracted,[62] provided that the solvent is chloroform. Irving et al.[63] were unable to prevent copper interference by using cobalticyanide, and they extracted copper and mercury at pH 1, in the absence of chloride, using chloroform. Copper could then be transferred to the aqueous phase by shaking with iodide at pH 4. Laug and Nelson[64] used bromide in acid solution to transfer mercury to the aqueous phase, leaving copper in the organic layer, which can be discarded. Vasak and Sedivek[65] extracted at higher pH, using thiocyanate to mask silver, and EDTA for other metals.

Absorptiometry of extracts may be either at 490 nm to measure absorbance of mercuric dithizonate in the presence of excess dithizone (the so-called mixed colour method) or at 620 to measure unreacted dithizone[66]. It should be mentioned that since solutions of dithizone and mercury dithizonate are photolabile, particularly in chloroform it is necessary to handle solutions in subdued light.[41]

It is possible to estimate mercury by titration with dithizone solution, by shaking the sample solution with successive portions until the persistence of the green colour shows that the reagent is in excess.

Extraction of mercuric dithizonate has been used, as is shown below, as a preliminary step in separating mercury for analysis by AAS or neutron-activation analysis.

c. Other colorimetric methods

Dinaphthyl carbazone forms a mercuric compound very similar to the dithizonate,[67] and methods for its general use[68] and for analysis of urine[69] and biological material[70] have been described. A specific test for mercury with glyoxal bis(2-thioanil) is described by Thabet and Tabibian.[71] Krasil'nikova et al.[72] give details of a method (used for mercury in metallic selenium) based on decolorisation by mercuric ion of copper diethyl dithio-carbamate in chloroform.

Several methods exploit the formation of complexes of mercuric-halide anions with dyes or other coloured compounds. Yamamoto et al.[73] extracted the orange-red complex of bromomercurate ion with tris 2, 2′-bipyridyl iron (III), using 1, 2-dichloroethane. Kolosova[74] found that a complex of iodo-mercurate with 2-(4-antipyrylazo-5-diethylaminophenol) could be extracted with several solvents to give blue-green solutions of molar absorptivity $1.6-2.1 \times 10^4$ at 600 nm, using the reaction for analysis of mercurial fungi-cides. Kothny[75] found that the complex of mercuric iodide and crystal violet could be extracted by toluene to give a sensitive ($\varepsilon = 56,000$ at 605 nm) reaction. Oshima and Nagasawa[76] observed that mercuric potassium iodide itself had moderately high absorbance ($\varepsilon = 15,800$ at 320 nm) and could also be used in a fluorometric determination. Tsubuchi[77] found that the complex of mercuric bromide and oxidised Bindschedler's green had high absorbance at 720 nm. Cherkesov et al.[78] report the use of compounds named Azoxin H and Azoxin C. Budensinsky and Svec[79] used glyoxal dithiosemicarbazone, of which the mercury compound has a molar absorptivity of 43,000 at 330 nm. Edrissi et al.[80] used the reaction of mercury (II) with tris (2-thiopyridine 1-oxide) iron (III), measuring the decolorisation of the reagent in chloroform at pH 4.0, but found it less sensitive than dithizone. Monnier and Gorgia[81] found that mercury forms a 1:1 complex with NADH and devised an absorptiometric method sensitive to 1 mcg Hg/ml and a fluorometric one sensitive to 0.03 mcg/ml.

d. Atomic absorption spectrometry

In these methods the absorption is measured, of radiation from a mercury vapour discharge (usually the line at 253.652 nm), by ground-state mercury atoms from the sample. The mercury atom population may be obtained by injecting a sample solution into a flame, or by generating mercury vapour either by thermal decomposition or chemical reduction of the sample or a concentrate. The method is inherently very sensitive[19] ($\varepsilon = 4.2 \times 10^6$), but in a given method the limit of detection will depend on the fraction of the available mercury which can be interposed (in the requisite ground state) in the radiation beam of the instrument. If a flame is used this fraction may be small, but flameless or cold-vapour methods can be devised in which virtually all the mercury available is effective.

Flame methods are practically specific for mercury, although some inter-ference by cobalt (which has a line at 253.649 nm) has been reported.[82] Cold-vapour methods, although free from metallic interference, may show errors if substances absorbent at 253.7 nm are present, such as solvent vapours, oxides of nitrogen or incompletely oxidised organic vapours or smoke. The effect can be overcome by separate measurement at wavelengths close to that of the mercury line.[83-85] Schachter,[84] and Jacobs,[86], in analysis of air, provided a reference path free of mercury by absorption in activated charcoal or an iodine solution. Barringer[87] designed an instrument in which the absorption at the centre part of the 253.7 nm line is corrected for absorbance not due to mercury by subtracting absorption from the wings of the pressure-broadened source. A somewhat similar system has been described by Ling.[88] Hadashei and McLaughlin[89] described an elegant system in which the Zeeman effect and a source containing ^{196}Hg are used to give two radiation beams, of which one measures total absorption and the other non-mercury absorption only.

Methods using direct injection into a flame are described by Lindstrom[90] and by Monkman et al.[91] (for urine, etc.). A concentrate by APDC/MIBK extraction was used by Willis[92] (urine), by Berman[93] and by Delaughter[94] (air). Messman and Smith[95] decomposed an APDC/MIBK extract from urine, by the boat technique. Dithizone extracts were used by Pyrih and Bisque[96] (soil and rock) and Devoto[97] (urine). Poluektov and Vitkun[98] used a flame technique and observed an increase in sensitivity when stannous chloride was added to the solution. This effect was confirmed by Hingle et al.[99]

Combustion in a stream of air or oxygen will dissociate mercury compounds, and the mercury vapour produced can be swept with the gas into the sample volume of an atomic absorption spectrometer. Methods are described for minerals,[100] colour additives,[101] soils, rocks and air,[87] organic materials,[85] and various other materials.[102]

The interpolation of gold or platinum coated on asbestos (or as a foil) in the gas stream traps mercury. This can subsequently be vaporised by heat, and measured without interference. Methods are given for soils and rocks,[103-105] and for air, blood, urine, fish, etc.[55,106-109]

An intermediate chemical separation of mercury may be used, such as concentration by APDC/MIBK[95] or dithizone extraction.[110-112] Monkman et al.[91] digested organic materials with acid and precipitated HgS, which was then ignited. Brandenberger and Bader,[113,114] Wheat,[115] and Doherty and Dorsett,[116] all used displacement or electrolysis to deposit mercury on copper wire or foil. Hinkle and Learned[117] and Fishman,[118] in water analysis, used silver. Toribara and Chields[119] digested tissues with HCl and NaNO$_3$, and concentrated mercury by ion exchange and co-precipitation with CdS which was then ignited.

Wet-ashing methods in which mercury vapour is generated by reduction are now common. Nearly as many digestion procedures as for dithizone methods have been described, as well as oxygen flask techniques.[120] The great sensitivity of the AAS determination allows use of samples of a few tenths of a gram, which simplifies the digestion and may make it more rapid. Most reduction procedures use $SnCl_2$ and NH_2OH at high acidity as described by Hatch and Ott.[121] Ascorbic acid is also effective.[122] In the Hatch and Ott procedure air is recirculated through the reduced solution, an $MgClO_4$ trap and the instrument cuvette, and the absorption increases as mercury vapour is swept out of the solution. It should reach a steady maximum, but in practice it declines, owing apparently to oxidation and absorption of the vapour. Uthe et al.[19] describe an apparatus which improves reproducibility. Their method, which uses low-temperature digestion intended primarily for analysis of fish, has been evaluated by Thorpe,[21] and has been adapted for partial automation.[33] A simple and precise method for transfer of mercury vapour, in another adaptation of this method, is described by Stainton.[20] Hoover et al.[123] have an adaptation of the Hatch and Ott method for analysis of foods, which are decomposed with HNO_3, and Krause et al.[124,148] give methods for urine, blood, water and air with H_2SO_4/HNO_3 digestion. Malaiyandi and Barrett[42] decomposed biological materials with $H_2SO_4/HNO_3/V_2O_5$. Osland[125] showed that iodide can interfere in the Hatch and Ott procedure but that most metals did not. Omang[126] has a method for water and effluents with $H_2SO_4/KMnO_4$ digestion, and Goulden and Afghan[127] one for water with destruction of organic compounds by ultraviolet photo-oxidation. Chau and Saitoh,[128] in water analysis, acidify and then concentrate the mercury by means of a dithizone extraction, followed by a back extract into a small volume. Lee and Lanfmann[129] show that cellulose materials can be digested with aqua regia. The Dow Chemical Company has published details of methods for water, brines, caustic, fish, sludges, mud, hydrogen, and air.[22]

Magos and Cernik[130] have developed a rapid method for determination of mercury in urine, without digestion. Dilute acid, cysteine and stannous chloride are added to the sample, through which air is drawn into the detector. No mercury vapour is released until excess sodium hydroxide is added. Organically bound mercury was not determined.

e. Neutron-activation methods

Methods using neutron-activation analysis can be made very sensitive and no chemical manipulation is required for the non-destructive methods. However, some interferences can be avoided by chemical separation of activated mercury after addition of carrier mercury. This separation need not be quantitative, provided it is reproducible, and contamination by inactive mercury after irradiation is of no importance. Access to both an atomic reactor with a flux of 10^{11}–10^{13} thermal neutrons/cm^2/sec and to counting equipment, usually with a pulse-height analyzer, is required.

When mercury of natural isotopic composition is irradiated with thermal neutrons five nuclides are produced but only ^{197}Hg of half-life 65 hr and ^{203}Hg of half-life 47 days are normally convenient for analytical use. Usually the 77 keV gamma rays and 68 keV X-rays from ^{197}Hg and the 279 keV gamma rays from ^{203}Hg are used for counting. The decay of ^{197}Hg, which is complex, is discussed in detail by Westermark and Sjostrand.[131] These authors found that losses of mercury by volatilization were considerable, and favoured a non-destructive procedure with irradiation and counting of samples sealed in quartz ampoules. The limit of sensitivity was about 0.03 mcg Hg. Smith,[132] by counting ^{203}Hg, of which the precursor ^{202}Hg is more abundant in natural mercury than ^{196}Hg, the precursor of ^{197}Hg obtained a limit of detection of 10^{-10} g. He irradiated biological material, added carrier mercury, digested with HNO_3/H_2SO_4, and after reduction of mercury to metal, separated it, redissolved, removed silver and gold as iodides, and precipitated the copper EDTA Hg_2I_2 complex for counting of ^{203}Hg.

Morris and Killick,[133] and Livingston et al.[134] counted ^{197}Hg after separation with iodide, copper and ethylene diamine. Sjostrand[135] and Hasanen[136] used the same radionuclide after electrolytic separation of mercury. Kim and Hoste[137] described a procedure for analysis of pure bismuth, and Kim and Stark[138] a method for urine. These workers used a rotating facility to get uniform flux during irradiation. Brune and Jirlow[139] found that samples caused considerable flux perturbation in the reactor. Brune[140] describes an arrangement for irradiation of frozen samples. Kamemoto and Yamaguchi[141] described a procedure in which only 0.1 mg carrier is added after irradiation. Mercury is separated as thiocyanate, ^{203}Hg is counted, and recovery is checked by reactivation and counting ^{197}Hg.

Non-destructive methods are described for blood, biological and environmental samples by Filby et al.,[142] for rocks and minerals by Laul et al.,[143] for petroleum by Shah et al.,[144] and for copper by Gillette.[145]

Digestion methods are given for biological materials,[136,146] for urine,[138,147] human hair,[124] plant tissue,[149,150] and rocks.[151,152] The use of ion exchange resins for separation of mercury in this analysis is described by Becknell et al.,[50] by Samsahl,[153] and by Jones et al.[154] Kosta and Byrne[155] ignited irradiated material in oxygen and trapped mercury in selenium-impregnated paper.

f. Miscellaneous methods

Winefordner and co-workers have shown[156-158] the feasibility of atomic fluorescence methods for mercury determination. Where the solution is injected into a flame detection down to about 0.1 ppm is possible.[157] By vaporising from a platinum loop into an atmosphere of argon about 10^{-8} g is detectable.[158] Vickers and Merrick[159] increased sensitivity by concentrating mercury by extraction with dithizone/CHCl$_3$ or APDC/MIBK. Vitkun *et al.*[160] used a low-temperature flame, and obtained sensitivities of around 2 mcg/ml, with SnCl$_2$ present.

An X-ray fluorescence method for mercury at concentrations of 2–40 ppm is described by Olsen and Shell.[161]

Emission spectrography, although not inherently of great sensitivity, was used for soils and rocks by Feges and Podobnik,[162] and after concentration of mercury on gold by Dall'Aglio.[163] Devices for controlling vaporisation in the arc source are described by Titov and Belobrov[164] and by Veres and Perfil'ev.[165]

Mass spectrometry has been used by Keith *et al.*[166] to identify mercury as the pollutant in a fish kill, and for quantitative determination by Tong *et al.*[167] and by Carter and Sites.[168]

Radiochemical methods depending on isotope exchange with Hg have been developed by Clarkson and Greenwood[169] (urine, tissues), by Magos and Clarkson[170] (air), and by Krivan.[171] An automated method using an exchange process followed by substoichiometric reaction with zinc dithizonate in CCl$_4$ and measurement of the activity of the organic phase has been described by Ruzicka and Lamm.[172]

Nascutiu[173] showed that by using K^{131}I, the complex [HgI$_4$] [CuA$_2$] (where A is ethylene or propylene diamine) could be prepared, and its activity used to measure the mercury concentration.

Hagiwara[174] describes a square-wave polarographic method of relatively low sensitivity and Perone and Kretlow[175] and Rannev *et al.*[176] described anode-stripping voltametric methods using a graphite electrode.

Methods of detection of mercury vapour by means of the stain produced on selenium-impregnated paper are given by Nordlander,[177] Stitt and Tomimatsu,[178] and Sergeant *et al.*,[179] and the operation of equipment using this paper is described.[180,181]

Pavlovic and Asperger[182] used the catalytic action of mercuric ions upon the reaction of ferrocyanide with nitrosobenzene, measuring the violet colour produced for determination of mercury in urine and blood. Krylova and Rubtsov[183] describe a microgravimetric method in which the complex of cupric iodide and mercuric iodide is precipitated. Thabet *et al.*[184] could detect 0.5 mcg Hg by reduction with hydroxylamine, with a ring-oven method, and Burton aud Irving[185] determined microgram amounts by a thermometric method in which oxidation of As(III) was catalysed by Hg.

There are a few methods useful for mercury in milligram amounts. They include a colorimetric method using Titan yellow[186] thermometric titration with periodate,[187] and a gravimetric procedure using 2-aminobenzothiazole as precipitant.[188]

Hansen[189] described a microbiological assay using Pseudomonas.

As an example of a flameless AAS method for total mercury, suitable for large numbers of samples, and for which the precision and comparability with other methods has been established, we think it justifiable to give some detail of a recent method.[33]

Samples of 0.1 to 0.5 g tissue are digested with a 1:4 mixture of HNO_3 (17N) and H_2SO_4 (36N) at 58C for 2 hr or until clear solutions are obtained. The samples are cooled in ice and excess $KMnO_4$ (6% w/v) is added. The flasks are stoppered, allowed to stand overnight, and H_2O_2 (30% w/w) is added dropwise until the precipitate of oxides of manganese just dissolves. The solutions are made to known volume. Mercury is reduced with a $NH_2OH/$ $H_2SO_4/SnSO_4$ mixture, and its vapour is equilibrated with air (using Technicon[R] Auto Analyzer[R] equipment) and determined by flameless AAS at a rate of 30 samples/hr. The absorbance of the vapour is registered on a strip-chart recorder and it is possible to set the instruments to give full chart width of 10 in with a mercury concentration in the test solution of 20 ng/ml. Blanks and standards are run with the solutions.

3. DETERMINATION OF INORGANIC AND ALKYL MERCURIALS

By far the greatest majority of quantitative methods for the measurement of mercury in biological materials are based on measurements of the mercury content after conversion of the various types of mercurials into mercuric ion. However, due to the differences in toxicological properties of various mercurials[6,25] it is necessary that good methods particularly for the alkyl-mercurials be available for determination of individual and groups of mercurials.

The earliest practical method for determining inorganic and organic mercurials in tissues was that of Gage[24] in 1961 and was based on the method of Polly and Millar[190] for the analysis of methylmercury dicyandiamide. These latter authors showed that methylmercury dicyandiamide in low concentration in water (down to 1 mcg/ml) could be easily converted into methylmercuric chloride which in turn was easily extracted into chloroform where it was measured spectrophotometrically as the dithizone complex. Gage found that organomercurials of the methyl- and phenylmercury type could be extracted into benzene from tissue homogenates in the presence of high concentrations of hydrochloric acid. Inorganic mercury did not partition into the benzene layer under these conditions. After this separation the organomercurials were extracted into aqueous sodium sulfide and the mercury content of this solution was determined by dithizone titration following permanganate oxidation. Recoveries of organic mercurials averaged 90% and levels of 1 mcg Hg/g could be determined. In 1969 Magos and Cernik[130] published a method for the direct determination of inorganic mercury in the presence of organic mercurials. The sample, either in liquid form or as a 1% sucrose homogenate was treated with acidic cysteine and stannous chloride. The sample was gassed for a short period and then 30% sodium hydroxide was added to liberate the bound mercury as elemental mercury. Further aeration of the sample volatilized the mercury and the amount of mercury was determined with the usual ultraviolet mercury detector fitted with a flow cell. The method was reported to be very rapid (one determination every 2 min) and required little in the way of specialized equipment. Organomercurial bonds were reported not to be cleaved by this procedure. The ability to gas the sample prior to release of the mercury without any loss of mercury from the solution allowed removal of trace volatile materials which might interfere if they were volatilized along with the mercury. In a later communication[191] Magos pointed out that the rate of release of mercury was dependent upon the biological nature of the sample being analyzed. For example, the rate of release from blood or kidney homogenate was essentially the same as that of saline solutions, but the rate of release from brain or fish homogenates was considerably reduced. This means that each type of sample must be standardized internally. Magos also pointed out that this procedure does not leave all carbon-mercury bonds intact. Although the reaction conditions are not vigorous enough to cleave the carbon-mercury bond in methylmercury, about one-third of the mercury is released from ethylmercuric chloride. As the author points out most environmental samples are contaminated with only methylmercury and therefore the method still gives a valid analysis of inorganic mercury. All mercurials could be broken down to mercuric ion by the addition of cadmium chloride along with the stannous ions. The rate of release again depends on the biological nature of the sample and it is necessary to carefully standardize conditions and use appropriate standardization procedures. With this procedure the average environmental sample and clinical experiment sample with methylmercury can easily be analyzed for their inorganic and total mercury content in a very short time.

In a like manner Umezaki and Iwamoto[192] studied the release of mercury from its various compounds in water by the action of stannous ion. They reported that in 2N sulfuric acid containing chloride ion mercury is formed only from inorganic mercury. No release of mercury from methylmercury was found under these conditions. However, in 1N sodium hydroxide solution containing trace amounts of cupric ion all of the mercury was released from methylmercury, ethylmercury, and phenylmercury as readily as it was released from mercuric compounds. This method was only applied to the analysis of water samples. The relative standard deviation of the analysis was 2% at the 5 mcg/1 level, using 100-ml samples.

None of the above methods distinguishes one organomercurial compound from another; only organic and inorganic groups are separated. In 1966

Japanese and Swedish workers each published methods for the determination of methylmercury in environmental samples. Both methods were based on the method of Gage for the extraction of methylmercuric chloride into organic solvents after treatment of the sample with strong halide acid. In the method of Kitamura *et al.*[193] methylmercuric chloride was extracted from the acidified sample into chloroform. Measurement of methylmercury was carried out using gas chromatography with electron capture detection. The detector responded to extremely minute amounts of methylmercury and the response (peak height) was linear up to 5×10^{-9} g methylmercuric chloride. Unfortunately in this procedure no detailed clean-up steps were used. The tissues were prewashed with water and acetone prior to extraction but the possibility of impure extracts interfering with the chromatographic procedure is great. Westöö's procedure,[15] on the other hand, employed a double partitioning clean-up procedure to remove interfering materials from methylmercury extracted from fish. Water homogenates of fish muscle, acidified with conc. HCl, were extracted with benzene. A portion of the benzene layer was removed and the volume was reduced for distillation after the addition of 0.1N acetic acid in heptane. The methylmercuric chloride was partitioned into ammonium hydroxide saturated with sodium sulfate. The methylmercuric hydroxide in a portion of the aqueous layer was then reconverted into methylmercuric chloride by the addition of hydrochloric acid and extracted into clean benzene. Gas chromatographic analysis was carried out on 10% Carbowax 20M on Chromosorb W, acid-washed, DCMS-treated (60–80 mesh) in a 128 cm × 3 mm stainless steel column. Quantitation of the amount of methylmercury present in the original sample was carried out by internal standardization, i.e. the sample was analyzed with and without the addition of a known amount of methylmercuric chloride. The use of ammonium hydroxide solution for extraction was acceptable only if prior distillation of the initial benzene extract was carried out. It was postulated that the distillation removed benzene-soluble thiols (e.g. methanethiol, hydrogen sulfide) which would prevent extraction of the methylmercury by the ammonium hydroxide due to formation of stable mercury-sulfide bonds. Westöö further stated that any methylmercury attached to a sulfur atom on a non-volatile compound which is not soluble in ammonium hydroxide will not be determined by this method. In a later paper[194] the methylmercury procedure was further extended to enable analyses to be carried out on a wide variety of foodstuffs and other materials. The original method, although giving good analysis on fish muscle, was not good for liver, eggs or meat, due to the formation of water-insoluble mercury-sulfur compounds upon treatment of the original benzene extract with ammonium hydroxide. Westöö found that these interfering sulfides could be blocked either prior to the first extraction with benzene by the addition of mercuric ions to the homogenate or by shaking the benzene layer with mercuric ions. Washing the initial benzene extract with mercuric chloride solution (5%) gave somewhat cleaner extracts but took longer than adding mercuric chloride to the homogenate prior to extraction with benzene. For extracts other than for fish and egg white an additional clean-up step was necessary prior to extraction with ammonium hydroxide. In these cases the benzene extract was quantitatively passed through an alumina column prior to distillation. Fish, egg and meat samples could also be analyzed by a much faster method. After preparation of the original benzene extract a portion was extracted with 1% aqueous cysteine saturated with sodium sulfate. After separation, a portion of the bottom layer was acidified, extracted, and analyzed as before. This latter procedure would not give satisfactory results, i.e. poor recoveries of added methylmercuric chloride were found for liver, aquaria sediments or sludges. In a final paper Westöö combined the above methods to give a method that has applicability to a wide variety of samples.[195] This procedure utilized the addition of mercuric chloride prior to extraction of the acidified homogenate with benzene and using aqueous cysteine rather than ammonium hydroxide for clean-up. Although this method gives good recoveries of added methylmercuric chloride with a wide variety of sample materials it suffers from the following disadvantage. Any dimethylmercury present in the original sample will react with mercuric ion, mole for mole, to give two moles of methylmercury. Thus the cysteine acetate clean-up without addition of mercuric chloride is still the preferred procedure, but its inability to handle certain samples must be kept clearly in mind. Westöö does this and modified this procedure in her 1968 paper[195] to include liver (treatment of original homogenate with molybdic acid) and egg yolk with low methylmercury content (use of multiple extractions with 1% cysteine acetate). She also recommended the use of glass gas chromatographic columns instead of the original stainless steel ones.

Westöö, in her original analyses of methylmercury in environmentally contaminated fish,[15] was careful to ensure qualitative identification of methylmercury. The clean-up benzene extracts of fish muscle on thin-layer chromatography gave spots with the same colour and Rf as authentic methylmercuric chloride. Removal of the spot followed by extraction and gas chromatography gave a peak with the same retention time as methylmercuric chloride. When the fish extracts and authentic methylmercuric chloride were treated to prepare the corresponding dithizonate, bromide, iodide or cyanide, chromatography, carried out as above, showed the fish extract and methylmercuric chloride to behave the same. Later the presence of methylmercury in benzene extracts was further confirmed by mass spectroscopy.[17]

During the period of Swedish development in this area, Japanese researchers were also continuing their work on analysis of organo-mercurials in environmental samples. Sumino[196] published a detailed study on the use of gas chromatography in this analysis. Sixteen different liquid phases were investigated and favourable results were obtained only with highly polar phases: DEGS (diethylglycol succinate), PEG-20M (polyethyleneglycol), and BDS (butanediol succinate). BDS in a light load (5% on Shimalite-W) and short (40 cm) glass column was recommended for analysis of aryl-mercurials. As little as 10^{-11} g methylmercuric chloride could be detected by electron capture detection, on well-aged columns, employing low oven temperatures. In practice, however, sensitivity such as this is rarely obtained due to column bleeding into the detector. Standard curves were determined with $(1-4) \times 10^{-10}$ g methylmercuric chloride. Detector response was linear over this range, but as is usual with electron capture detectors standard curves had to be determined frequently. Methylmercuric chloride, iodide, hydroxide, acetate, and sulfate all elute with about the same retention time. This is thought to be due to the conversion of all of these compounds to a common anion form of methylmercury.[197]

Extraction procedures for determining the organomercurials in a variety of materials were described. Biological materials were homogenized with water, then washed with acetone and ethyl acetate. After removal of the solvent the methylmercury residue was extracted with 1N HCl. The methylmercury chloride was then extracted into benzene for analysis. The above procedure was only applicable to materials containing relatively high levels of mercury, for example, shellfish from Minamata Bay.

For analysis of material containing less methylmercury a procedure very similar to the Swedish methods was adopted. Acidified aqueous fish homogenates were extracted with benzene three times. The benzene layer was then extracted three times with aqueous glutathione. The aqueous layer, now containing the methylmercury as a glutathione derivative, was then acidified and the methylmercury was partitioned into benzene for analysis. Even with this clean-up Sumino[196] reports that certain materials yielded final extracts which were still contaminated with substances which interfered with the subsequent gas chromatographic determination. In these cases additional clean-up of the sample was performed by thin-layer chromatography on silica gel. Yamaguchi and Matsumoto[198] reported further studies on the gas chromatographic behaviour of alkylmercury compounds. In confirmation of earlier work BDS on Shimalite-W was the preferred support. Electron capture detection was preferred to any other method and the authors report marked increase in sensitivity in switching from a standing DC voltage to pulse DC collection. (Nitrogen as a carrier gas gave the best response.) On-column injection with glass columns also increased detector response. Dilute solutions of methylmercuric chloride in benzene could be concentrated by heating the solution and aspirating the air overlying the surface. Recoveries of methyl-mercury averaged 90%. No information is presented on equivalent studies with other evaporators, e.g. flash evaporator Kuderna-Danish, N_2 stream. Westöö[15] states that dilute solutions of methylmercuric chloride in benzene can be concentrated by distillation at atmospheric pressure by adding a few millilitres of acidic heptane prior to distillation. Yamaguchi and Matsumoto also reported that high levels of mercuric chloride (200 ppm) in ether give a response that could be misidentified as methylmercuric chloride. Teramoto et al.[199] repeated this finding but reported that the interfering material was removed by washing with aqueous glutathione. The presence of interfering materials in inorganic mercurials has been confirmed by other workers.[194,197]

Jensen[200] summarized some of the problems associated with the determination of methyl mercury and in particular problems associated with interfering materials in the final benzene extract. He suggested that after analysis of the final benzene solution for methylmercury the solution should be shaken with aqueous $AgSO_4$ and the benzene phase reanalyzed. If the peak was truly methylmercuric chloride no peak should be obtained after this treatment, as any methylmercuric chloride present would have been converted into methylmercuric sulfate and partitioned into the aqueous phase. Jensen also studied the presence of materials in inorganic mercurials which chromatographed as methylmercury. This material was analyzed by a combination of gas chromatography-mass spectroscopy. The peak eluting from the chromatograph as methylmercury chloride gave no evidence of methylmercury ion on mass spectroscopic analysis. The presence of methylmercury ion can be confirmed by chemical confirmation according to the following reaction.

$$CH_3HgBr + CH_2N_2 \rightarrow CH_3HgCH_2Br$$

The methylmercury methylene bromide formed could be analyzed for by gas chromatography. It eluted with retention time different from that of methylmercuric bromide.

In 1970 the Swedish Water and Air Pollution Research Laboratory released their method[201] for the determination of methylmercury in fish. The procedure is generally similar to the previously mentioned Swedish and Japanese methods. The methylmercury present in fish is converted to the halide form by the use of strong acid and extracted into an organic phase. The extract was then cleaned up by partitioning the methylmercury into an aqueous phase and then back into an organic phase. The chemical reagents used in these procedures were different from those used previously. Methylmercury was extracted from tissue by the combined action of copper and bromide ions in the presence of strong H_2SO_4. The methylmercuric bromide is extracted from the toluene with dilute aqueous sodium thiosulfate. After the addition of 3M KI the methylmercuric iodide formed was extracted with benzene and analyzed by gas chromatography.

Gas chromatographic analysis was carried out at 140°C on 7% Carbowax 20M on Varaport-30 (100/200 mesh), using N_2 as a carrier gas. The sensitivity of the method is reported to be 10^{-11} g and samples containing 0.5 mcg/kg can be analyzed. Recoveries of added methylmercury after correcting for loss of toluene (only a portion of the original 10 ml of toluene was taken for analysis) were approximately 90%.

This procedure was considerably modified by Uthe's group.[202] Extraction and partitioning of the methylmercuric bromide into toluene was carried out in a single step employing ball mills of the type described by Grussendorf et al.[203] In place of the steel ball bearings used in the original equipment glass marbles were used. After centrifugation, 5 ml of the toluene layer was extracted twice with 0.005M $Na_2S_2O_3$—95% ethanol (1:1, v/v). The use of ethanol in this step prevented formation of emulsions. The methylmercury thiosulfate complex was detroyed by the addition of potassium iodide. The methylmercuric iodide was then extracted into benzene for analysis by gas chromatography. All of the clean-up partitioning steps were carried out in small glass vials designed for fast partitioning and use of small volumes.[202] Gas chromatographic analysis was carried out on 7% Carbowax 20M on high-performance Chromosorb W (80/100 mesh) using nitrogen as a carrier gas at 170°C. Recoveries of methylmercury added to the sample prior to grinding averaged 99 ± 5%. In their original procedures only one extraction with ethanolic thiosulfate was used and this samples was analyzed by the use of internal standardization. Standard curves were determined using methylmercuric chloride rather than methylmercuric iodide due to the lability of methylmercuric iodide. The detector response (peak height) was identical for either compound with respect to the amount of methylmercury injected. The detector used was a Varian concentric tube detector fitted with a tritium foil and operated in a standing D.C. mode. The authors recommended the use of this detector because of the ease with which it can be cleaned and returned to operation as compared to many other electron capture detectors. Even though the operation of the detector was in the standing D.C. mode rather than pulsed D.C. the sensitivity of the chromatographic system was such that 10^{-11} g of methylmercuric chloride could be detected. Inorganic and elemental mercury did not interfere with methylmercury analysis and dimethylmercury was only 1–2% changed to methylmercury by this treatment. This illustrated the advantage of using copper ion over mercury ion to block interfering sulfydryl groups. The modifications and use of specialized equipment markedly reduced the man-hour requirement. For example, the use of the ball mills not only effectively prevents solvent loss but also allows centrifugation to break emulsions. Use of small amounts of solvents also adds a tremendous health and fire-safety factor.

Even after this clean-up certain extracts still were able to adversely affect chromatographic behaviour and required further clean-up. This was accomplished by passing the final benzene extract through a small column of florisil (0.5 g) which was overlaid with Na_2SO_4 (~0.5 g). No significant loss of methylmercuric iodide was found at this step (95–100% recovery).

Fujiki[204] studied the effectiveness of a few metal ions in liberating methylmercury from sulfide compounds that could not be broken down by strong halide acids, e.g. bis-dimethyl mercury sulfide to yield methylmercuric ion. Mercuric chloride, cuprous chloride and silver chloride effectively released methylmercury. Sodium, calcium, magnesium, zinc, and ferrous chlorides were not effective. Silver chloride, however, was not recommended since it had an adverse effect on the gas chromatographic part of the procedure. The presence of material in mercuric chloride which acted like methylmercury chloride was noted and thus cuprous chloride was recommended. Nishi *et al.*[205] also studied the effect of various metal ions on the release of methylmercury from these sulfur compounds. Mercuric chloride was found to be the most effective but cupric chloride could be used.

Tatton and Wagstaff[206] pointed out that none of the earlier methods is suitable for the determination of alkoxyalkyl mercurials since they are usually acid-labile. The method they described was successfully applied to apples, tomatoes and potatoes. Samples were extracted three times with isopropanol and 1% cysteine hydrochloride. The combined extract was clarified by centrifugation and then diluted with aqueous Na_2SO_4 and the resulting solution was extracted three times with diethylether. After discarding of the organic layer the mercurials were extracted with three portions of 0.005% dithizone in diethylether. After drying with sodium sulfate and concentration (Kuderna–Danish evaporator) the mercurials were analyzed by gas chromatography. Recoveries of 85–95% were reported for methyl-, ethyl-, and ethoxyethylmercuric chlorides at 0.01 to 1 mcg/g levels and 0.5 to 5 mcg levels for phenyl- and tolymercuric acetates.

Hartung[207] presented a method for the analysis of both methylmercury and dimethylmercury in the same samples. Sample material was homogenized with water containing sodium tetraborate and cysteine hydrochloride. The homogenate was then extracted three times with toluene. After separation of the final toluene phase the residual aqueous phase was acidified with an equal volume of conc. HCl and extracted three more times with toluene to extract the methylmercury. The original pooled toluene extract containing the dimethylmercury was saponified for 2 hr with KOH. After removal of the aqueous layer the toluene layer was washed twice with water. Emulsions formed during these washings were broken with NaCl. After drying over

Na_2SO_4 the dimethylmercury in the toluene was converted to methylmercuric bromide by refluxing with an aqueous solution of $HgBr_2$ and KBr. After clean-up and drying the methylmercury was determined as below.

The toluene phase containing methylmercuric chloride from the sample was extracted with fresh cysteineborate solution. The lower layer was then acidified with conc. HCl as before and the methylmercuric chloride was analyzed by gas chromatography on 11% QF1+OV17 on Gas-chrom Q(80/100 mesh) at 100°C, using nitrogen as a carrier gas. The conversion of dimethylmercury to mono-methylmercury was essentially that of Jensen and Jernelov.[18] It is similar to the method of Westöö,[195] except that she used mercuric chloride in dilute HCl. Studies using methylmercuric chloride and dimethyl mercury showed 80–95% of the added mercury was recovered.

Newsome[208] in 1971 reported that attempts to utilize the methods of either Westöö[195] or Sumino[196] were unsatisfactory with fish of high fat content. Emulsion formation at the initial extraction stage and low recoveries of added methylmercury were the major problem. Neither method was at all satisfactory for analysis of cereal grain products. Two procedures, one for fish and the other for cereal products, were published.

Recoveries of added methylmercury averaged $96 \pm 6\%$ for fish, $99 \pm 6\%$ for oats, and $105 \pm 11\%$ for wheat flour. Fish were extracted twice with 2.1N KBr in 1.0N HBr. The combined extracts were extracted with benzene. Any emulsions remaining in the organic layers after removal of the aqueous layer were broken by shaking with solid anhydrous sodium sulfate. The combined benzene extracts were extracted with cysteine acetate. A portion of the cysteine layer was acidified with HBr and the methylmercury bromide was then extracted into benzene. The HBr had to be exhaustively extracted with benzene prior to use to remove interfering material.

Cereal products (10 g) were ground for 5 min with benzene: 90% formic acid (10:1, v/v). A portion of the filtrate was layered on to a silicic acid column. The column was run at approx. 1 ml per min. with benzene and the fraction eluting between 15–55 ml was collected. The eluate was shaken with cysteine acetate solution and after separation of the layers a portion of the cysteine acetate solution was acidified with 48% HBr and the methylmercuric bromide was extracted into 1 ml benzene. Gas chromatographic analysis was carried out on 2% BDS on Chromosorb W, acid-washed, and DMCS-treated (100/200 mesh), using nitrogen as a carrier gas at 120°C. Methylmercuric chloride was used to prepare standard curves. As reported earlier the response was equivalent with respect to methylmercury, regardless of the halide anion.

The quantitative analysis of sediments and waters for methylmercury has not been developed to the degree to which it has been developed for biological materials. Natural levels of total mercury in water vary from 0.02 to

0.7 ppb in fresh water and 0.03 to 2 ppb in sea water. Hosohara *et al.*[209] found between 1.6 and 3.6 ppb in waters from Minamata Bay. In general, fresh waters contain 0.01 to 0.2 ppb[210-212] and fresh water in areas of known deposits of mercury were somewhat higher.[211,213] Nishi and Horimoto[214] studied the extraction of alkylmercury chloride from water. Water acidified to 0.1N with HCl was then extracted twice with benzene. The benzene extract was then extracted with a small volume of 0.1% cysteine. After acidification of the cysteine layer with HCl the methylmercuric chloride was extracted into benzene. As little as 0.41 mcg of methylmercuric chloride per litre was successfully determined. Considering the levels of total mercury in water noted above, a method at least 100 times as sensitive as the above procedure is needed. Nishi *et al.*[199] studied the use of selective ion exchange resins for concentration of methylmercuric chloride from water. Cotton cellulose containing 3.4% sulfhydryl was prepared by reacting cotton with thioglycolic acid and acetic acid in the presence of sulfuric acid and acetic anhydride. As little as 0.1 g of this resin could retain 0.1 mcg of methylmercuric chloride from one litre of water. The methylmercuric chloride was then eluted with 2N HCl. The eluant was extracted with 1 ml of benzene for analysis. Recoveries at this level were slightly in excess of 100%. No results were reported for methylmercury compounds of sulfur, i.e. the usual anion for methylmercury in nature. Also, with environmental samples containing many other chemicals besides methylmercury, the efficiency of the resin could be considerably impaired. Nishi pointed out that this resin could pick up materials such as inorganic mercury and cadmium as well.

The analysis of sediments has been studied to a limited degree in both Sweden and Japan. In general the problem in soils is not one of low concentration, since normal soils and sediments have total mercury levels of 0.02 to 0.15 mcg/g[215] and thus are in the same concentration range as non-contaminated fish.

Westöö reported[195] that her original method could not be applied to sediment analysis due to the presence of interfering non-volatile sulfur compounds of methylmercury. The inability of either ammonium hydroxide or cysteine acetate to form water-soluble methylmercuric derivatives was given as the reason. She found that pretreatment of the sediments with aqueous mercuric chloride was necessary to give good recoveries of methylmercuric chloride. Nishi studied the use of mercuric chloride in this regard and found that it was the most effective agent for removal of interfering sulfur compounds. According to Nishi[199] it was, however, most important that residual mercuric chloride be removed from the reaction mixture prior to acidification and extraction of methylmercuric chloride. He reported that failure to do this resulted in mercuric chloride being carried through extraction and clean-up and giving erroneous high values for methylmercury chloride on gas chromatographic analysis. Excess mercury was removed by the addition of ammonium hydroxide. After removal of the precipitate the methylmercuric chloride was determined in the usual manner.

Nishi further reported that either cuprous or cupric ion could be used in place of mercury but their efficiency was somewhat reduced. Gillespie,[216] using the procedure of Uthe, found that even with sample sediments that had been spiked and allowed to air-dry prior to analysis, over 90% of the added methyl mercuric chloride could be recovered. Although dimethylmercury was not hydrolyzed during analysis of fish tissue[202] the situation with regard to hydrolysis during analysis of sediments is not known. The procedure of Hartung[207] has not to our knowledge been applied to material other than tissue.

The problems associated with determination of levels of organomercurial compounds in air are much like those associated with the determination of organomercurials in water. The normal total mercury concentration in air is reported to be of the order of $10^{-8} - 10^{-9}$ g per cubic metre. Air over known mercury deposits can run as high as 10^{-6} g per cubic metre and use of this has been made in identifying new mercury-containing deposits.[4] Goldwater[217] found between $(1-41) \times 10^{-6}$ g per cubic metre for indoor air samples and between 0 and 10^{-8} g per cubic metre for outdoor samples. Samples collected over busy superhighways[218] ran as high as 18×10^{-6} g per cubic metre, whereas air over fields where organomercurial fungicides had been used were of the order of 10×10^{-6} g per cubic metre. A value of 10^{-5} g per cubic metre was recommended as a threshold limit in 1965,[219] and it is at this level that work on measuring air levels of organomercurials has been carried out.

Kimura and Miller[220] successfully collected methyl- and ethylmercuric chloride as well as mercury vapour from dilute solution in air. The organomercurials were absorbed in two absorbers in series each containing aqueous sodium carbonate and disodium phosphate. Mercury vapour was absorbed in the third absorber by acid permanganate. Recovery studies were carried out by aspirating air above dilute solutions of organomercurials. The flow rate was about 200 litres per hr. Under these conditions 100 to 1000 mcg of either methyl- or ethylmercuric chloride could be collected with an efficiency. approaching 100%. Only trace (0-2%) amounts of the organomercurial were found in the second absorber. Mercury vapour was not absorbed in either of these absorbers but was quantitatively picked up in the acid permanganate absorber.

Christie et al[221] reported that organomercurials in air at concentrations in the region of 10^{-5} g per cubic metre could be selectively absorbed by either glass fiber pads impregnated with cadmium sulfide or by a fluidized bed of activated charcoal. The cadmium sulfide method could absorb ethylmercuric chloride, ethylmercuric phosphate,. diphenylmercury and methylmercury dicyandiamide with good recovery. With this method only 20% of diethyl-mercury was retained. The fluidized bed method successfully recovered all of the above-mentioned compounds. Total levels of organomercurials ranged between 5 and 20×10^{-6} g per cubic metre.

Jensen and Jernelov[18] demonstrated the presence of dimethylmercury in air by bubbling the air through toluene. The toluene was shaken with mercuric chloride (10 mg) and KBr (110 mg) to yield two moles of methylmercuric bromide per mole of dimethylmercury. Unfortunately no data on the ability of toluene to extract dimethylmercury from air were presented.

4. GAS CHROMATOGRAPHIC CONSIDERATIONS

Quantitative gas chromatographic analysis of dilute methylmercuric halide and other organomercurial materials in organic solvents is generally carried out using electron capture detection systems, although some work has been done with flame ionization detection and thermoconductivity bridges,[196] but these were abandoned in favour of electron capture detectors due to the extreme sensitivity of these detectors. Recently[222] the use of emission spectrometry in a microwave powered inert gas plasma as a detector has been applied to gas chromatography as a detection system, with a sensitivity approaching that of electron capture detection. This detector is also reported to have a high specificity for mercury. The electron capture detection system has advantages associated with both extreme sensitivity and a certain degree of selectivity. By working with a reasonable degree of care and quality of equipment, detection of 10^{-11} g of methylmercuric chloride is readily attainable and measurement of levels of contamination down to $10^{-8} - 10^{-9}$ g/g appear feasible with one-gram samples. This sensitivity of the detector, however, puts a few basic requirements on the gas chromatographic systems.

The electron capture detector works by measuring the decrease in current flow across a potential.[223] Electrons for current flow are generated by low energy β^- particles. As an electron capturing material flows through the detector the number of electrons available for current flow decreases and generates the signal. For maximum sensitivity it is essential that very little electron capturing materials be present in the eluant from the chromatographic column. This generally means the use of high-purity carrier gases and low-bleed column packing materials, such as the silicone greases. Another problem which arises is the plating of materials on to the ion source surface. As the thickness of this plating increases, the amount of electrons available for capture is decreased and the detector functions less and less efficiently. There are at present a number of ways of minimizing the plating of ion source surface. High-temperature detectors (utilizing ^{63}Ni as a β^- source) are usually not as seriously contaminated this way, due to the increased volatility of the foreign materials at these higher temperatures. In detectors operated in a standing DC mode, material is constantly plated on to the β^- source electrically. Use of pulsed DC collection (e.g. a 3 microsecond pulse with 5–150 microsecond interval) lowers the amount of electroplating. At the same time, use of pulsed detection is generally a more sensitive system, as the pulse interval allows for a period of electron build-up in the detector, which in turn allows for an increased current flow. At the same time the potential for electron capture is much greater, so that the response of the detector to an electron capturing material is increased. Another method of minimizing plating effects is simply one of detector design. Detectors such as the Varian (Walnut Creek, Calif.) concentric tube electron capture detector are designed in such a way that it is very easy to remove and clean the β^- source. Although many instrument manufacturers recommend cleaning of the foil with solvents (such as mild alcoholic KOH), the use of a mild abrasive,[224] such as chrome polish, has been found to be much more effective and capable of removing little ^{3}H from the foil.

The use of very sensitive detection systems which are themselves very sensitive to contamination demand that samples presented to the instrument for quantitation be as free from contaminating and interfering materials as possible. Materials present in extracts may not only impair detector operation but might also interfere with quantitative detection in a number of other ways. Examples have been found of material which is electron capturing and elutes at the same time as methylmercury, thus leading to an apparent increase in the methylmercury content. Other materials can affect the chromatographic column in such a manner that methylmercury decomposes and either does not give a methylmercury response or gives a repressed response. Materials can also impair the chromatographic efficiency of the column, resulting in less sharp peaks being eluted. Care must be taken to ensure adequate performance at the system and in most laboratories utilizing these systems standard curves (injection of known amounts of methylmercury) are run and detector characteristics (column efficiencies, peak height or area, peak skewness) are determined frequently. Fortunately these detection systems have found widespread use in regulatory laboratories concerned with determining residual amounts of pesticide in foodstuffs and other materials. Many detailed and excellent manuals[225-227] are available from these groups and it is strongly recommended that anyone unfamiliar with pesticide residue analysis consult these manuals prior to attempting methylmercury determinations. Some of the earlier work done on gas chromatography of organomercurials employed metallic columns and indirect column injection. These often lead to breakdown of methylmercury during chromatography and loss of sensitivity, so that most workers have now switched to the following sort of model system. The gas chromatograph should be set up for direct on-column injection fitted with glass columns. Low-bleed polar liquid phases such as DEGS,[199] Carbowax 20M,[15] BDS polyethylene-glycol succinate,[206] QF1–OV17[207] coated on high-efficiency supports such as Chromosorb W,[36] Shimalite,[196] Chromosorb W, acid-washed, HMDS-treated,[195,208] Gas-Chrom Q[207] at 2–25% loading have been employed as chromatographic agents. Nitrogen was widely used as a carrier gas and oven operating temperatures range from 100°C to 175°C as columns get longer and supports more polar. In general, columns are conditioned by gas purging for extended periods at high temperatures prior to use. Although column performance deterioration can be kept to a minimum by using care to ensure that only cleaned-up samples are injected, a few other "tricks" have been used to maintain and improve chromatographic performance. Westöö[195,228] states that new columns can be improved by injection of benzene solutions of organic (1 mg/ml) and inorganic (aqueous $HgCl_2$; benzene extract) prior to use. Nishi et al.[199] reported that column efficiency can be increased and tailing eliminated by impregnation of the solid support (Chromosorb W) with 5–10% sodium chloride or potassium bromide prior to application of the liquid phase (DEGS). Using these columns he found by mass spectrometry that regardless of the anion of the methylmercury injected the eluted methylmercury was in the anionic form of the halide used to coat the column. Uthe et al.[202] found that periodic injection (5 mcl) of 3M aqueous KI sufficed to ensure good column operation. Hartung[207] reported that periodic heating of the column (11% QF−1+OV17 on Gas-Chrom Q 80/100 mesh) to 210°C to clear out interfering materials with long retention times resulted in good column operation.[207]

Extraction and clean-up of samples appears feasible by most of the above-mentioned procedures. They all give recoveries of added methylmercury, which indicates that clean-up procedures do not result in unacceptable losses from fish. With other materials the situation is not as straightforward and it is advisable to proceed with caution. The actual situation with respect to fish is not as clear-cut as it appears.

Westöö,[229] having compared directly her own methods with that of Sumino,[196] reported that Sumino's method gave lower values than her earlier method.[195] She believed that the lower values obtained by the Japanese method were due to the use of insufficient acid during extraction which led to incomplete extraction of the methylmercury into the benzene phase. Also, she found that the amount of glutathione required to extract all of the methylmercuric chloride from the benzene layer was greater than that used by Sumino.

In 1968 Japanese and Swedish workers collaborated in a check sample study by exchanging fish samples. The results[230] showed that on the average the Japanese values were significantly lower than the Swedish methylmercury levels (63–94% with a mean of 77%). Similar studies were carried out in Sweden. These studies showed 80% of the reported values were within $\pm 10\%$ of the mean value and all values were within $\pm 20\%$ of the mean. Currently, laboratories in Sweden, Japan, Canada, and the United States are involved in a detailed comparative sample program for methylmercury in fish[231] and it is to be hoped that in a year or two an internationally recognized method for the determination of methylmercury will be available.

5. COMPARATIVE STUDIES ON TOTAL MERCURY ANALYSIS TECHNIQUES

The situation regarding analytical studies of methods for determination of the total mercury content is much better than that for methylmercury or organomercury methods. The recommended method[23] of the AOAC is based on that of Klein[232] and was shown to yield satisfactory results in a check sample study involving a number of laboratories.[232] The method of Laug[233] performed as well but was not selected due to the rigorous reagent purifications necessary.

Interlaboratory comparisons involving both destructive and non-destructive neutron activation were carried out[234] using two flour samples, one of which had a natural mercury level of about 0.5 mcg/g, the other was spiked to a level of about 5 mcg/g. Although the amount of non-destructive neutron-activation analysis was very limited there was no gross difference in the results obtained by destructive or non-destructive analysis. Out of 17 laboratories one laboratory had results grossly different from the mean with respect to the treated flour, but five laboratories had results with the untreated flour which differed grossly from the mean values. The result of the results on the treated flour (mean = 4.59 ± 1.32 mcg/g) lay between 2.5 mcg/g and 6.5 mcg/g, whereas those on the untreated (mean = 0.044 ± 0.014 mcg/g) flour lay between 0.03 mcg/g and 0.06 mcg/g. A small intercomparative study on mercury levels was carried out in Japan in 1967,[235] using human hair samples. On the average, neutron-activation results were 8–20% higher than dithizone values and differences between two different laboratories ranged as high as 100%. In a comparison between Swedish neutron activation and Japanese atomic absorption analysis,[230] using both Japanese and Swedish fish samples, Japanese values on Swedish fish were somewhat higher (109–123%) than the Swedish values, whereas Japanese values on Japanese fish were only 64–82% of the Swedish values.

No significant difference was found between neutron-activation methods and atomic absorption methods as applied to fish analysis in a check sample study by Uthe et al.[235] Disregarding laboratories which reported results which differed markedly from the mean the 25 other laboratories had the following gross means: A = 1.34 ± 0.26; B = 0.12 ± 0.10; C = 4.12 ± 0.84.

Flame atomic absorption results were just significantly lower than the results from neutron-activation or flameless atomic absorption methods. No difference was found between results from two Canadian laboratories, one employing flameless atomic absorption, the other the dithizone method, in a small intercomparative study.[236] A detailed comparison between flameless atomic absorption and the AOAC method[237] was carried out by Munns and Holland.[238] These workers reported that the atomic absorption methods gave more consistent results and better recoveries of added mercury than the AOAC method. Results from the AOAC method were erratic and recoveries ranged between 0 and 106%. These must be compared to the good recoveries reported by Klein.[232] Recoveries of added mercury by flameless atomic absorption ranged between 76 and 92%. Uthe[236] in a study of flameless atomic absorption found recoveries of greater than 90% when spikes ranging from $(1–4) \times 10^{-9}$ g of either mercuric chloride or methylmercuric chloride were added to fish samples (0.1–0.3 g) prior to digestion. Munns and Holland[238] also reported that increasing quantities of fish digest in the final solution lowered the recovery of mercury. We have looked for such an effect, i.e. if the effect exists for all atomic absorption methods, analysis of smaller portions of fish should give higher mercury values than if larger amounts were weighed out. This has never been found in our experience. These authors also recommended the use of molybdate in the digestion procedure and it is possible that the molybdate has a significant effect on the final portion of the analysis.

In summary it may be stated that there are at least three good general methods for measuring mercury in biological tissues at a parts-per-million level. Neutron activation, however, requires extremely expensive facilities, highly trained personnel and lengthy procedures. Colorimetric analysis is also good but requires expensive glassware, good chemists and is also lengthy. Atomic absorption analysis, on the other hand, does not require expensive equipment, as even the cheapest of mercury meters can be used. Highly trained chemists are not required as there are no highly skilled procedures involved other than sample weighing. The method is also easily adapted to a large daily output of analyses.[19,33]

6. BIBLIOGRAPHY

1. G. Wobeser, N. O. Nielson, R. H. Dunlop, and F. M. Atton, *J. Fish. Res. Bd. Can.* **4**, 830 (1970).
2. E. G. Bligh, *Mercury and the Contamination of Freshwater Fish*. Fish. Res. Bd. Can. Manuscript Rep. Ser. No. 1088, Winnipeg, Man., April (1970).
3. D. McAlpine and S. Araki, *Lancet* **2**, 629 (1958).
4. F. D'Itri, *The Environmental Mercury Problem*. Technical Report No. 12 (Institute of Water Research, Mich. State Univ., (1971).

5. H. Harada, Congenital or Fetal Minamata Disease. In *Minamata Disease*. Study Group of Minamata Disease (Kumamoto University, Japan, 1918), pp. 93–117.
6. P. L. Bidstrup, *Toxicity of Mercury and Its Compounds* (Elsevier, New York, 1964).
7. T. Takeuchi, *Biological Reactions and Pathological Changes of Human Beings and Animals under the Condition of Organic Mercury Contamination*, presented at the International Conference on Environmental Mercury Contamination, Ann Arbor, Mich., Sept. 30–Oct. 2 (1970).
8. K. Irukayama, *Advan. Water Poll. Res.* **3**, 153 (1967).
9. T. L. Kurland, S. N. Faro, and H. Seidler, *World Neurol.* **1**, 370 (1960).
10. K. Sumino, *Kobe J. Med. Sci.* **14**, 131 (1968).
11. T. Tsubaki, T. Sato, and K. Kondo, *Japan Med. J.* **6**, 132 (1967).
12. K. Borg, *Sveriges Natur* **51**, 92 (1960).
13. K. Borg, H. Wanntorp, K. Erne, and E. Hando, *J. Appl. Ecol. Suppl.* **3**, 171 (1966).
14. T. Westermark, *The Mercury Problem in Sweden*, 1964. Natural Resource Study. Stockholm, Sweden (1965).
15. G. Westöö, *Acta Chem. Scand.* **20**, 2131 (1966).
16. K. Noren and G. Westöö, *Var Föda* **2**, 13 (1967).
17. G. Westöö, In *Chemical Fallout*, edited by M. W. Miller and G. G. Berg (Charles C. Thomas Publ., Springfield, Ill., 1969), p. 75.
18. S. Jensen and A. Jernelov, *Nature* **223**, 753 (1969).
19. J. F. Uthe, F. A. J. Armstrong, and M. P. Stainton, *J. Fish. Res. Bd. Can.* **27**, 805 (1970).
20. M. P. Stainton, *Anal. Chem.* **43**, 625 (1971).
21. V. A. Thorpe, *J. Ass. Offic. Agr. Chem.* **54**, 206 (1971).
22. Anon., *Determination of Mercury by Atomic Absorption Spectrophotometric Method— Applicable to: Water, Brines, Caustic, Fish, Sludges, Mud, Hydrogen and Air* (Dow Chemical Company, Midland, Mich., 1970).
23. Committee on Editing Methods of Analysis, *Official Methods of Analysis of the Association of Official Agricultural Chemists* (Ass. Offic. Agr. Chem. Washington, D.C., 1965), 10th ed.
24. J. C. Gage, *Analyst* **86**, 457 (1961).
25. G. Lofroth, *Methylmercury—a Review of Health Hazards and Side Effects Associated with the Emission of Mercury Compounds into Natural Systems*. Working Group on Environmental Toxicology, Ecological Research Committee of the Swedish Natural Science Research Council (Stockholm, Sweden 1969).
26. L. Fishbein, *Chrom. Rev.* **13**, 83 (1970).
27. N. A. Smart, *Residue Rev.* **23**, 1 (1968).
28. T. T. Gorsuch, *Analyst* **84**, 135 (1959).
29. G. F. Smith, *Anal. Chim. Acta.* **8**, 397 (1953).
30. H. E. Monk, *Analyst* **86**, 608 (1961).
31. W. Horwitz (Ed.), *Official Methods of Analysis of the Association of Official Agricultural Chemists* (Washington, D.C., 1970), 11th ed.
32. M. B. Jacobs, S. Yamaguchi, L. J. Goldwater, and H. Gilbert, *Amer. Ind. Hyg. Ass. J.* **21**, 475 (1960).
33. F. A. J. Armstrong and J. F. Uthe, *Atomic Abs. Newsl.* **10**, 101 (1971).
34. H. Barnes, *J. Mar. Biol. Ass. U.K.* **26**, 303 (1946).
35. M. B. Jacobs and A. Singerman, *J. Lab. Clin. Med.* **59**, 871 (1962).
36. F. N. Kudsk, *Scand. J. Clin. Lab. Invest.* **16**, 575 (1964).
37. J. A. Pickard and J. T. Martin, *J. Sci. Food Agr.* **11**, 374 (1960).
38. F. N. Ward and J. B. McHugh, *U.S. Geol. Survey Prof. Paper No.* 501-D, 128 (1964).

39. E. A. Epps, *J. Ass. Offic. Anal. Chem.* **49,** 793 (1966).
40. N. A. Smart and A. R. C. Hill, *Analyst* **94,** 143 (1969).
41. F. M. Kunze, *J. Ass. Offic. Anal. Chem.* **31,** 438 (1948).
42. M. Malaiyandi and J. P. Barrette, *Anal. Lett.* **3,** 579 (1971).
43. D. C. Abbott and E. I. Johnson, *Analyst* **82,** 206 (1957).
44. A. C. Rolfe, F. R. Russell, and N. T. Wilkinson, *Analyst* **80,** 523 (1955).
45. H. Thies, *Deut. Apoth. Ztg* **106,** 193 (1966). In *Anal. Abstr.* **14,** 3667 (1967).
46. D. M. Goldberg and A. D. Clarke, *J. Clin. Path.* **23,** 178 (1970).
47. L. Palalau, *Rev. Chim.* (Bucharest) **20,** 107 (1969). In *Chem. Abstr.* **69,** 3329u (1968).
48. J. F. Reith and C. P. van Dijk, *Chem. Weekbl.* **37,** 186 (1940).
49. R. Liebmann and D. Hempel, *Fortschr. Wasserchem. Ihrer Grenzeb.* 173 (1969). In *Chem. Abstr.* **71,** 7380t (1969).
50. D. E. Becknell, R. H. Marsh, and W. Allie, *Anal. Chem.* **43,** 1230 (1971).
51. E. G. Rochow, D. T. Hurd, and R. N. Lewis, *The Chemistry of Organometallic Compounds* (Wiley and Sons, New York, 1953).
52. W. H. Gutenmann and D. J. Lisk, *J. Agr. Food Chem.* **8,** 306 (1960).
53. M. Fujita, Y. Takeda, T. Terai, O. Hoshino, and T. Ukita, *Anal. Chem.* **40,** 2042 (1968).
54. M. N. White and D. J. List, *J. Ass. Offic. Anal. Chem.* **53,** 530 (1970).
55. V. Lidums and U. Ulfvarson, *Acta. Chem. Scand.* **22,** 2379 (1968).
56. T. Kato, S. Takei, and A. Okagami, *Japan Analyst* **5,** 689 (1956).
57. A. M. Kiwan and H. M. N. H. Irving, *Anal. Chim. Acta.* **54,** 351 (1971).
58. Y. Kurayaki and K. Kusumoto, *Japan Analyst* **16,** 815 (1967).
59. F. Koroleff, *Meerentutkimuslait. Julk.* **145,** 1 (1950).
60. F. J. Welcher, *Organic Analytical Reagents* (Van Nostrand, New York, 1946).
61. W. O. Winkler, *J. Ass. Offic. Agr. Chem.* **21,** 220 (1938).
62. R. F. Milton and J. L. Hoskins, *Analyst* **72,** 6 (1947).
63. H. Irving, G. Andrew, and E. J. Risdon, *J. Chem. Soc.* 541 (1949).
64. E. P. Laug and K. W. Nelson, *J. Ass. Offic. Anal. Chem.* **25,** 399 (1942).
65. V. Vasak and V. Sedivek, *Chem. Listy* **45,** 10 (1951). In *Chem. Abstr.* **45,** 6532c (1951).
66. E. B. Sandell, *Colorimetric metal analysis* (Interscience, New York, 1959).
67. I. B. Supronovich, *J. Gen. Chem. USSR* **8,** 839 (1936).
68. F. W. Laird and A. Smith, *Ind. Eng. Chem.* (*Anal. Ed.*) **10,** 576 (1938).
69. D. M. Hubbard, *Ind. Eng. Chem.* (*Anal. Ed.*) **12,** 768 (1940).
70. J. Cholak and D. M. Hubbard, *Ind. Eng. Chem.* (*Anal. Ed.*) **1,** 149 (1946).
71. S. Thabet and O. Tabibian, *Anal. Chim. Acta* **34,** 241 (1966).
72. L. N. Krasil'nikova, L. I. Maksai, and M. N. Chepik, *Sb. Nauch. Tr. Vses. Nauch-Issled. Gornomet. Inst. Tset. Metal.* 23 (1968). In *Chem. Abstr.* **72,** 1817m (1970).
73. Y. Yamamoto, S. Kikuchi, Y. Hayashi, and T. Kamanura, *Japan Analyst* **16,** 931 (1967).
74. I. V. Kolosova, *Izv. Vyssh. Ucheb. Zaved. Khim.* **12,** 1239 (1969). In *Chem. Abstr.* **72,** 96358 (1970).
75. E. L. Kothny, *Amer. Ind. Hyg. Ass. J.* **31,** 466 (1970).
76. G. Oshima and K. Nagasawa, *Eisei Kagaku* **16,** 78 (1970).
77. M. Tsubuchi, *Anal. Chem.* **42,** 1087 (1970).
78. A. I. Cherkesov, V. S. Toukoshkurov, A. I. Postoronko, and V. N. Ryshov, *Zh. Anal. Khim* **25,** 466 (1970).
79. B. W. Budesinsky and J. Svec, *Anal. Chim. Acta* **55,** 115 (1971).
80. M. Edrissi, A. Massoumi, and J. A. W. Dalziel, *Microchem. J.* **15,** 579 (1970).
81. D. W. Monnier and A. Gorgia, *Anal. Chim. Acta* **54,** 505 (1971).

82. D. C. Manning and F. Fernandez, *Atomic Abs. Newsl.* **7**, 24 (1968).
83. A. E. Ballard, D. W. Stewart, W. O. Kamm, and C. W. Zuelke, *Anal. Chem.* **26**, 921 (1954).
84. M. M. Schachter, *J. Ass. Offic. Anal. Chem.* **49**, 778 (1966).
85. R. Woodriff and D. Schrader, *Anal. Chem.* **43**, 1918 (1971).
86. M. B. Jacobs and R. Jacobs, *Amer. Ind. Hyg. Ass. J.* **26**, 261 (1965).
87. A. R. Barringer, *Trans. Inst. Min. Metall. B* **75**, 120 (1966).
88. C. Ling, *Anal. Chem.* **40**, 1876 (1968).
89. T. Hadashei and R. D. McLaughlin, *Science* **174**, 404 (1971).
90. O. Lindstrom, *Anal. Chem.* **31**, 461 (1959).
91. J. L. Monkman, P. A. Maffett, and T. F. Doherty, *Amer. Ind. Hyg. Ass. J.* **17**, 418 (1956).
92. J. B. Willis, *Anal. Chem.* **34**, 615 (1962).
93. E. Berman, *Atomic Abs. Newsl.* **6**, 57 (1967).
94. B. Delaughter, *Atomic Abs. Newsl.* **9**, 49 (1970).
95. B. B. Messman and B. S. Smith, *Atomic Abs. Newsl.* **9**, 81 (1970).
96. R. Z. Pyrih and R. E. Bisque, *Econ. Geol.* **64**, 825 (1969).
97. G. Devoto, *Rass. Med. Sarda*, Suppl. **71**, 289 (1968). In *Chem. Abstr.* **73**, 1127x (1970).
98. N. S. Poluektov and R. A. Vitkun, *Zh. Anal. Khim.* **18**, 37 (1963). In *Chem. Abstr.* **59** 1083b (1963).
99. D. N. Hingle, G. F. Kirkbright, and T. S. West, *Analyst* **92**, 759 (1967).
100. Y. N. Kusnetsov and L. P. Chabovskii, *Uch. Zap. Tsentr. Nauchn.-Issled. Inst. Olovyan. Prom.* 75 (1964). In *Chem. Abstr.* **63**, 4936f (1965).
101. J. A. Wenninger, *J. Ass. Offic. Anal. Chem.* **48**, 265 (1965).
102. V. A. Razumov and T. P. Utkina, *Spektrosk. Tr. Sib. Soveshch.* **1065**, 291 (1969). In *Chem. Abstr.* **74**, 19050r (1971).
103. C. H. James and J. S. Webb, *Trans. Inst. Min. Metall.* **73**, 633 (1963).
104. W. W. Vaughn and J. H. McCarthey, *U.S. Geol. Survey Prof. Paper* No. 501-D, 123 (1964).
105. G. Friedrich and M. Kulms, *Erzmetall.* **22**, 74 (1969). In *Chem. Abstr.* **72**, 7443c (1970).
106. S. H. Williston, *J. Geophys. Res.* **73**, 7051 (1968).
107. G. Thilliez, *Chim. Anal.* (Paris) **50**, 226 (1968).
108. D. H. Anderson, J. F. Evans, J. J. Murphy, and W. W. White, *Anal. Chem.* **43**, 1511 (1971).
109. G. W. Kalb, *Atomic Abs. Newsl.* **9**, 84 (1970).
110. U. Ulfvarson, *Acta Chem. Scand.* **21**, 641 (1967).
111. F. N. Kudsk, *Scand. J. Clin. Lab. Invest.* **17**, 171 (1965).
112. K. S. Shater, A. F. Eremina, A. I. Semeryakova, V. V. Smirnov, and A. V. Efremov. *Zavod. Lab.* **36**, 1470 (1970). In *Chem. Abstr.* **74**, 57127d (1971).
113. H. Brandenberger and H. Bader, *Atomic Abs. Newsl.* **6**, 101 (1967).
114. H. Brandenberger and H. Bader, *Atomic Abs. Newsl.* **7**, 53 (1968).
115. J. A. Wheat, *Rep. Atom. Energy Comm. U.S.D.P.* 1164, 10 pp. (1968).
116. P. E. Doherty and R. S. Dorsett, *Anal. Chem.* **43**, 1887 (1971).
117. M. E. Hinkle and R. E. Learned, *U.S. Geol. Survey Prof. Paper* No. 650-D, 251 (1969).
118. M. J. Fishman, *Anal. Chem.* **42**, 1462 (1970).
119. T. Y. Toribara and C. P. Chields, *Amer. Ind. Hyg. Ass. J.* **29**, 87 (1968).
120. E. G. Pappas and L. A. Rosenberg, *J. Ass. Offic. Agr. Chem.* 49, 782 (1966).
121. W. R. Hatch and W. L. Ott, *Anal. Chem.* 40, 2085 (1968).
122. S. Shimomura, Y. Fukamoto, M. Hashimoto, and Y. Tanase, *Bunseki Kagaku* **19**, 1296 (1970).

123. W. L. Hooever, J. R. Melton, and P. A. Howard, *J. Ass. Offic. Agr. Chem.* **54**, 860 (1971).

124. L. A. Krause, R. Henderson, H. P. Shotwell, and D. A. Culp, *Amer. Ind. Hyg. Ass. J.* **32**, 331 (1971).

125. R. Osland, *Spectrovision* **24**, 11 (1970).

126. S. H. Omang, *Anal. Chim. Acta* **53**, 415 (1971).

127. P. D. Goulden and B. K. Afghan, *Technical Bull.* No. 27 (Inland Water Branch. Department of Energy, Mines and Resources, Ottawa, Ont., 1970).

128. Y. K. Chau and H. Saitoh, *Environ. Sci. Technol.* **4**, 489 (1970).

129. D. C. Lee and C. W. Laufmann, *Anal. Chem.* **43**, 1127 (1971).

130. L. Magos and A. A. Cernik, *Brit. J. Ind. Med.* **26**, 144 (1969).

131. T. Westermark and B. Sjostrand, *Int. J. Appl. Radiat. Isotopes* **9**, 1 (1960).

132. H. Smith, *Anal. Chem.* **35**, 635 (1963).

133. D. F. C. Morris and R. A. Killick, *Talanta* **11**, 781 (1964).

134. H. D. Livingston, H. Smith, and N. Stovanovic, *Talanta* **14**, 505 (1967).

135. B. Sjostrand, *Anal. Chem.* **36**, 814 (1964).

136. E. Hasanen, *Suomen Kemistilehti*, B **43**, 251 (1970). In *Chem. Abstr.* **73**, 106199u (1970).

137. J. I. Kim and J. Hoste, *Anal. Chim. Acta* **35**, 61 (1966).

138. J. I. Kim and H. Stark, *Radiochim. Acta* **14**, 213 (1970).

139. D. Brune and K. Jirlow, *Radiochim. Acta* **8**, 161 (1967).

140. D. Brune, *Acta Chem. Scand.* **20**, 1200 (1966).

141. Y. Kamemoto and S. Yamaguchi, *Nature* **202**, 487 (1964).

142. R. H. Filby, A. I. Davis, K. R. Shah, and W. A. Haller. Mikrochim. Acta 1130 (1970).

143. J. C. Laul, D. R. Case, M. Wechter, F. Schmidt-Bleek, and M. E. Lipschutz, *J. Radioanal. Chem.* **4**, 241 (1970).

144. K. R. Shah, R. H. Filby, and W. A. Haller, *J. Radioanal. Chem.* **6**, 412 (1970).

145. R. K. Gillette, *Anal. Lett.* **4**, 563 (1971).

146. K. K. S. Pillai, C. C. Thomas, J. A. Sonda, and C. M. Hyche, *Anal. Chem.* **43**, 1419 (1971).

147. K. Ishida, S. Kawamura, and M. Izawa, *Anal. Chim. Acta* **50**, 351 (1970).

148. H. Kawakami, *Onken Kiyo* **20**, 218 (1968). In *Chem. Abstr.* **72**, 64862b (1970).

149. W. A. Haller, L. A. Ranatelli, and J. A. Cooper, *J. Agr. Food Chem.* **16**, 1036 (1968).

150. C. M. Silva and F. W. Lima, *Revta. Bras. Tecnol.* 67 (1970). In *Anal. Abstr.* **21**, 2971 (1971).

151. A. Alian and R. Shabana, *Microchem. J.* **12**, 427 (1967).

152. W. Kiesl, Nat. Bur. Stand. (U.S.) Spec. Publ. (1969) 302 pp.

153. K. Samsahl, *Anal. Chem.* **39**, 1380 (1967).

154. J. D. Jones, J. M. Rottschafer, H. B. Mark, K. E. Paulsen, and G. J. Patriarche, *Mikrochim. Acta* 399 (1971).

155. L. Kosta and A. R. Byrne, *Talanta* **16**, 1297 (1969).

156. J. D. Winefordner and R. A. Staab, *Anal. Chem.* **36**, 165 (1964).

157. J. M. Mansfield, J. D. Winefordner, and C. Veillon, *Anal. Chem.* **37**, 1049 (1965).

158. M. P. Bratzel, R. M. Dagnall, and J. D. Winefordner, *Anal. Chim. Acta* **48**, 197 (1969).

159. T. J. Vickers and S. P. Merrick, *Talanta* **15**, 873 (1968).

160. R. A. Vitkun, N. S. Poluektov, and Y. V. Zelyukova, *Zh. Anal. Khim.* **25**, 474 (1970). In *Chem. Abstr.* **73**, 21071w (1970).

161. E. C. Olsen and J. W. Shell, *Anal. Chim. Acta* **23**, 219 (1960).

162. J. Feges and B. Podobnik, *Vestn. Slov. Khem. Drus.* **16**, 5 (1969). In *Chem. Abstr.* **72**, 106788q (1970).

163. M. Dall'Aglio, *Atti Sci. Soc. Toscona Sci. Nat. Mem. Ser. A.* **73**, 553 (1966). In *Chem. Abstr.* **68**, 6060v (1968).

164. E. V. Titrov and V. M. Belobrov, *Zavod. Lab.* **36**, 248 (1970).

165. G. I. Veres and A. P. Perfil'ev, *Zavod. Lab.* **36**, 331 (1970).

166. L. H. Keith, A. W. Garrison, M. M. Walker, and A. L. Alford, *Amer. Chem. Soc. Div. Water Air Waste Chem. Gen. Papers* **9**, 3 (1969).

167. S. S. C. Tong, W. H. Gutenmann, and D. J. Lisk, *Anal. Chem.* **41**, 1872 (1969).

168. J. A. Carter and J. R. Sites, *Anal. Lett.* **4**, 351 (1971).

169. T. W. Clarkson and M. P. Greenwood, *Talanta* **15**, 547 (1968).

170. L. Magos and T. W. Clarkson, *Brit. Pat.* **1**, 130921 (1970).

171. V. Krivan, *Z. Anal. Chem.* **253**, 192 (1971).

172. J. Ruzicka and C. G. Lamm, *Talanta* **15**, 689 (1968).

173. T. Nascutiu, *Acad. Rep. Popolare. Romine Studii Ceretari chim.* **8**, 649 (1970).

174. K. Hagiwara, *Osaka Kogyo Gijuta Shikenjo Kiho* **19**, 213 (1968). In *Chem. Abstr.* **71**, 18542d (1969).

175. S. P. Perone and W. J. Kretlow, *Anal. Chem.* **37**, 968 (1965).

176. G. G. Rannev, V. S. Volkova, and A. M. Murtazaei, *Zavod. Lab.* **36**, 1446 (1970). In *Chem. Abstr.* **21**, 2448 (1971).

177. B. W. Nordlander, *Ind. Eng. Chem. (Anal. Ed.)* **19**, 518 (1927).

178. F. Stitt and Y. Tomimatsu, *Anal. Chem.* **23**, 1098 (1951).

179. G. A. Sergeant, B. E. Dixon, and R. G. Lidzey, *Analyst* **82**, 27 (1957).

180. Anon., *Methods for the detection of Toxic Substances in Air*, Booklet No. 13, *Mercury* (HMSO, London, 1957).

181. Anon., *A Handbook of Colorimetric Chemical Analytical Methods*, Part 6: *The Determination of mercury in air*. (The Tintometer Ltd., Salisbury, England, 1952).

182. D. Pavlovic and S. Asperger, *Anal. Chem.* **31**, 939 (1959).

183. A. N. Krylova and A. F. Rubtsov, *Vopr. Sudebnoi Med., Min. Zhravookhr. U.S.S.R.* 328 (1968). In *Chem. Abstr.* **72**, 87102t (1970).

184. S. K. Thabet, N. E. Salibi, and P. W. West, *Anal. Chim. Acta* **49**, 575 (1970).

185. K. C. Burton and H. M. N. H. Irving, *Anal. Chim. Acta* **52**, 441 (1970).

186. D. Negoiu and A. Kriza, *Analele Univ. "C. I. Parhon" Bucuresti, Ser. Stiiat. Nat. Chim.* **11**, 117 (1962). In *Chem. Abstr.* **60**, 13870a (1964).

187. B. M. Tejam and B. C. Halder, *Ind. J. Chem.* **7**, 613 (1969).

188. S. N. Tandon, P. K. Srivasta, and S. R. Joshi, *Mikrochim. Acta* 1208 (1970).

189. J. C. Hansen, *Nord. Hyg. Tidskr.* **51** (1970). In *Chem. Abstr.* **73**, 86627b (1970).

190. D. Polly and V. Millar, *J. Agr. Food Chem.* **2**, 1030 (1954).

191. L. Magos, *Analyst* **96**, 847 (1971).

192. Y. Umezaki and I. S. Iwamoto, *Japan Analyst* **20**, 173 (1971).

193. S. Kitamura, T. Tsukamoto, K. Hayakawa, K. Sumino, and T. Shibata, *Igaku to Seibutsugaku* **72**, 274 (1966).

194. G. Westöö, *Acta Chem. Scand.* **21**, 1790 (1967).

195. G. Westöö, *Acta Chem. Scand.* **22**, 2277 (1968).

196. K. Sumino, *Kobe J. Med. Sci.* **14**, 115 (1968).

197. S. Nishi, Y. Horimoto, and R. Kobayashi, *Identification and Determination of Trace Amounts of Organic Mercury*, presented at the International Symposium for Identification and Measurement of Environmental Pollutants, Ottawa, Ont., June (1971).

198. S. Yamaguchi and H. Matsumoto, *Kurume Med. J.* **16**, 33 (1969).

199. R. Teramoto, M. Kitabatake, M. Tanabe, and Y. Noguchi, *J. Chem. Soc.* (Japan) **70**, 1601 (1967).

200. S. Jensen, *Nord. Hyg. Tidskr.* **50**, 85 (1969)

201. H. O. Bouveng, *Determination of Methylmercury by Gas Chromatography* (Swedish Water and Air Pollution Laboratory Publication C7A, Stockholm, Sweden, 1968).

202. J. F. Uthe, J. Solomon, and B. Grift, *J. Ass. Offic. Anal. Chem.*, **55** 583 (1972)

203. O. W. Grussendorf, A. J. McGinnis, and J. Solomon, *J. Ass. Offic. Anal. Chem.* **53**, 1048 (1970).

204. M. Fujiki, *Japan Analyst* **19**, 1507 (1970).

205. S. Nishi, Y. Horimoto, and Y. Umezawa, *Japan Analyst* **19**, 1646 (1970).

206. J. O. G. Tatton and P. J. Wagstaffe, *J. Chromatogr.* **44**, 284 284 (1969).

207. R. Hartung, *The Determination of Mono- and Dimethylmercury Compounds by Gas Chromatography*, presented at the Conference on Environmental Mercury Contamination, Ann Arbor, Mich., Sept. 30–Oct. 2 (1970).

208. W. H. Newsome, *J. Agr. Food Chem.* **19**, 567 (1971).

209. K. Hosohara, H. Kozuma, K. Kawasaki, and Isuruta, *Nippon Kagaku Zasshi* **82**, 1979 (1961). In *Chem. Abstr.* **56**, 576 (1962).

210. L. Wiklander, *Geodeuma* **3**, 75 (1969). In *Chem. Abstr.* **71**, 6391n (1969).

211. M. Pall'Anglio, In *Origin and Distribution of the Elements* edited by L. H. Ahrens (Pergamon Press, Oxford, 1968) p. 1065.

212. A. Stock and F. Cucuel, *Naturwissenschaften* **22**, 390 (1934).

213. D. Brune and D. Landstrom, *Radiochim. Acta* **5**, 228 (1966).

214. S. Nishi and Y. Horimoto, *Japan Analyst* **17**, 75 (1968).

215. G. I. Mekhonina, *Pochvovedenie* **11**, 116 (1969). In *Chem. Abstr.* **72**, 110364r (1970).

216. D. C. Gillespie, Personal communication.

217. L. J. Goldwater, *J. Roy. Inst. Public Health Hyg.* **27**, 279 (1964).

218. Y. Fujimura, *Japan J. Hyg.* **18**, 10 (1964).

219. *Threshold Limit Values* 1965, adopted at the American Conference of Governmental Hygienists, 27th Annual Meeting, May (1965).

220. Y. Kimura and V. L. Miller, *Anal. Chem.* **32**, 420 (1960).

221. A. A. Christie, A. J. Dunsdon, and B. S. Marshall, *Analyst* **92**, 185 (1967).

222. C. A. Bache and D. J. Lisk, *Anal. Chem.* **43**, 950 (1971).

223. *Aerograph Research Notes*, Winter Issue (Wilkens Instruments, Walnut Creek, Calif., 1965).

224. A. V. Holden and G. A. Wheatley, *J. Gas. Chromatogr.* **5**, 373 (1967).

225. H. P. Burchfield and D. E. Johnson, *Guide to the Analysis of Pesticide Residues* (U.S. Dept. of Health, Education and Welfare, Office of Pesticides, Washington, D.C., 1965), 5 Volumes.

226. J. A. Burke, J. E. Gaul, and P. E. Corneliussen (Eds.), *Pesticide Analytical Manual* (U.S. Dept. of Health, Education and Welfare, Food and Drug Administration, Washington, D.C., 1971), 3 Volumes.

227. H. A. MacLeod, P. J. Wales, R. A. Graham, M. Osadchuk, and N. Blumen, *Analytical Methods for Pesticide Residues in Foods* (Food and Drug Directorate, Ottawa, Ont., 1969).

228. L. Kamps and B. Malone, *Determination of Methylmercury in Fish and Animal Tissues by Gas-Liquid Chromatography*. Preprint copy (Food and Drug Administration, Washington, D.C.).

229. G. Westöö. *Comparison Between Japanese and Swedish Methods for Determining Methylmercury Compounds in Fish Flesh* (National Institute of Public Health, S-10401 Stockholm 60), Stencils 1968.

230. Anon., *Methylmercury in Fish* (Uno S. Andersons Tryckeri, Stockholm, 1971).

231. D. M. Smith, Personal communication.

232. A. K. Klein, *J. Ass. Offic. Anal. Chem.* **35**, 537 (1952).

233. J. Laug, *J. Ass. Offic. Anal. Chem.* **25,** 399 (1942).
234. A. Tugsaval, D. Merten and O. Suschnv, *The Reliability of Low-level Radiochemical Analysis.* Results of Intercomparisons Organized by the Agency during the Period 1966–1969 (IAEA, Vienna, March 30, 1970). Stencils.
235. I. Kawasaka, H. Tanabe, Y. Hosomi, T. Kondo, M. Takeda, K. Kanoda, T. Tatsono, T. Amano, K. Matsui, T. Kobata, C. Ukita, O. Hoshino, K. Tanzawa, K. Ueda, M. Nishimuro, H. Aoki, T. Kondo, Y. Nakazawa, and T. Ishikura. State of pollution due to mercury compounds, Part II, In *Report on the Cases of Mercury Poisoning in Niigata* (Ministry of Health and Welfare, Tokyo, 1967) Stencils, pp. 75–265.
236. J. F. Uthe, F. A. J. Armstrong, and K. C. Tam, *J. Ass. Offic. Agr., Chem.* **54,** 866 (1971).
237. J. F. Uthe, F. A. J. Armstrong, and K. C. Tam, *Paper presented at the Fifth Pesticide Residue Analysts Seminar*, Vancouver, B.C., May (1970).
238. R. K. Munns and D. C. Holland, *J. Ass. Offic. Agr. Chem.* **54,** 202 (1971).

Mass Spectra of Some Fungicidal Organomercury Compounds[†]

O. HUTZINGER, W. D. JAMIESON, and S. SAFE

Atlantic Regional Laboratory, National Research Council of Canada, Halifax, Nova Scotia, Canada

Mass spectra of ten organomercury compounds are reported and their major fragmentation pathways presented. Application of mass spectrometry to the identification of such compounds in natural samples is discussed and high resolution data are applied to the identification of methylmercuric dicyandiamide (Panogen) in treated seeds.

INTRODUCTION

There is great concern, currently, over the finding of organomercury compounds in the environment, particularly in biological samples from wildlife and fish, and about their possible toxicological effects.[1−3]

In many cases these compounds (mainly methylmercury derivatives[2,4,5]) appear to be formed by biological reaction from inorganic mercury derivatives released into the environment as industrial waste. A structurally diverse group of organomercury derivatives, however, has been applied on to agricultural land as seed dressings, turf fungicides and for other fungicidal uses.[6,7]

The present methodology of analysis for organomercury compounds includes very sensitive procedures (see Refs. 6, 8, 9; see also Refs. 10–13); however, in all these methods no differentiation between various compounds is made since only a "total mercury" value is obtained. Procedures involving GLC and TLC (see Refs. 6, 8 and 9; see also 14–16) differentiate between various compounds, and preliminary characterization is possible through comparison of R_f-values and retention times.

† Issued as NRCC No. 12120

55

Mass spectrometry of organomercury compounds offers the advantage of unambiguously identifying small samples directly. Dimethylmercury and methylmercuric halides, for example, have been identified in fish extracts by mass spectrometry.[4,17] Samples can be purified by GLC or TLC. Because of the presence of six significantly abundant Hg isotopes (^{198}Hg, 10.02%; ^{199}Hg, 16.84%; ^{200}Hg, 23.13%; ^{201}Hg, 13.22%; ^{202}Hg, 29.80%; ^{204}Hg, 6.85%), ions in the spectra containing Hg are recognized easily and identification of crude or partially purified samples is possible. Identification of very small samples or low concentrations is further aided by the mass deficiency of the mercury isotopes (cf. identification of organochlorine pesticides[18]).

The purpose of this paper is: (i) to indicate the possibility of unambiguously identifying Hg compounds by mass spectrometry involving only simple clean-up steps, (ii) to provide reference spectra of some of the most commonly used organomercury fungicides and related compounds, and (iii) to show the mechanism of electron-impact-induced fragmentation of these compounds.

Surprisingly little work has been done on the behavior of organomercury compounds on electron impact and only data on dialkylmercury[19–22] and some pentafluorophenylmercury derivatives[23] are available.

MATERIALS AND METHODS

Compounds

Mercury derivatives used and their source are listed in Table I. All were analytical-grade samples and most spectra were recorded on the samples as received. "Thimerosal" (**10**) was converted from the sodium salt to the free acid by treatment with dilute acid, ether extraction and recrystallization from toluene. Phenylmercuric salicylate (**9**), methoxyethylmercuric chloride (**5**) and ethoxyethylmercuric chloride (**6**) were purified by preparative TLC (**9**, ether:benzene:acetic acid, 10:10:0.3, $R_f = 0.6$; **5**, hexane:acetone, 3:1, R_f 0.38; **6**, hexane:acetone, 3:1, R_f 0.45).

Phenylmercuric acetate (phenyl-d_5) was prepared by direct mercuration of benzene (d_6) with mercuric acetate.[24] Phenylmercuric acetate (**2**), propionate (**3**), salicylate (**9**), and chloride (**4**) were heated at 150° for 20 min, whereby compounds **1–3** melted. The residue was shaken with benzene and the benzene solution spotted on silica plates and chromatographically compared to the starting materials and diphenylmercury in solvent A (hexane) and B (hexane:acetone, 3:1).

Extraction of Organomercury Fungicides from Seeds

Sample A   Barley (100 g) treated with Panogen PX (methylmercury dicyandiamide 0.9%) at 2 fl. oz. per bushel was extracted with ethanol (3×150 ml).

TABLE I
Organomercury compounds investigated and mass spectral conditions

| | | | | Temperature (°C) | |
| | | | | Ion source | Sample probe |
Number	Name	Formula	Source[a]	Ion source	Sample probe
1	Diphenyl-mercury	$C_6H_5HgC_6H_5$	B	167	54
2	Phenylmercuric acetate	$C_6H_5HgOCOCH_3$	A	165	55
3	Phenylmercuric propionate	$C_6H_5HgOCOCH_2CH_3$	A	162	52
4	Phenylmercury chloride	$C_6H_5Hg—Cl$	C	169	70
5	Methoxyethyl-mercuric chloride	$CH_3OCH_2CH_2—Hg—Cl$	B	166	28
6	Ethoxyethyl-mercuric chloride	$CH_3CH_2OCH_2CH_2—Hg—Cl$	B	164	27
7	Methylmercuric acetate	$CH_3HgOCOCH_3$	D	164	26
8	Methylmercuric dicyandi-amide	$CH_3—Hg—NHCNHCN$, with $\overset{NH}{\overset{\|}{}}$	F	188	—[b]
9	Phenylmercuric salicylate	$C_6H_5HgO—CO—C_6H_4—o\text{-}OH$	B	174	—[b]
10	Ethylmercuri-thiosalicylic acid	$HOOC\text{-}o—C_6H_4—S—Hg—CH_2CH_3$	E	166	89

[a]Source B Alfa Inorganics   D Strem Chemicals Inc.
C Aldrich Chemical Co.   E Elanco Products (Eli Lily & Co.)
A Chemical Procurement Labs.   F Pesticide Repository, Perrine Primate Research Branch, F.D.A.
[b]Samples recorded with the standard probe. The actual sample temperature would be ca. 5° below the ion source temperature.

The crude extract was filtered and a sample was transferred directly into the mass spectrometer sample tube after evaporation of the solvent.

Sample B   Wheat (200 g) treated with Pandrinox liquid fungicide–insecticide (1.33 oz. methylmercury dicyandiamide per imperial gallon) at 0.902 ml per lb. was extracted with benzene (3 × 250 ml). The extract was prepared for mass spectrometry in the same way as sample A.

Mass Spectrometry

The 70-volt mass spectra were obtained with a DuPont/C.E.C. model 21-110B mass spectrometer. The samples were introduced directly into the source, using either a standard probe or a probe which allowed the temperature of the sample to be controlled independently from the source temperature.[25] High-resolution spectra were recorded on photoplates (Ilford Q2); perfluoro-kerosene (PFK) was used as an internal mass standard. The plates were exposed for 30 min to the ions from the crude samples A and B after the probe had attained the temperature indicated in Table I.

RESULTS AND DISCUSSION

Fragmentation

All spectra are characterized by a group of peaks at m/e 198–204 ($Hg^{+\cdot}$). Because of the large number of mercury isotopes, "metastable peaks" were quite diffuse.

The electron-impact-induced fragmentation pathway for diphenylmercury (**1**) is shown in Figure 1. The ions **a–e** are also present in the mass spectra of the substituted phenylmercury compounds **2**, **3**, and **9** due to their facile

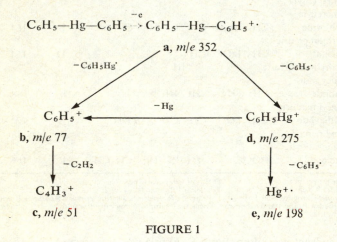

FIGURE 1

thermal conversion into diphenylmercury itself. This reaction can also be demonstrated by heating **2**, **3** and **9** for 20 min at 150° and reexamination of the products by TLC, which confirms the formation of diphenylmercury (R_f in solvent A, 0.15; in solvent B, 0.8). Phenylmercuric chloride (**4**), acetate (**2**), and propionate (**3**) fragment by initial loss of chlorine, acetate, and propionate from their respective molecular ions to give the $C_6H_5Hg^+$ ion (**d**). The mass spectrum of the chloride (**4**) shows also a weak but characteristic $HgCl^+$ ion at m/e 233; the acetate (**2**) yields a strong acetic acid peak at m/e 60 and the propionate (**3**) a propionic acid peak at m/e 74. In the mass spectra of

$C_6D_5HgOCOCH_3$ the ion at m/e 60 is replaced by a peak at m/e 61 indicating H (or D) transfer from the aromatic ring (Figure 2).

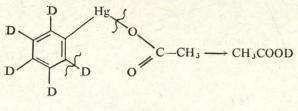

FIGURE 2

The acetate (2) and propionate (3) also show the expected M—CH_3 (m/e 319) and M—C_2H_5 (m/e 319) ions respectively resulting from cleavage α to the carboxyl group. The latter compound expels CO_2 to give an ion at m/e 304.

The spectra of phenylmercuric salicylate (9) also shows the ions a–e but strong characteristic peaks at m/e 138, m/e 120 and m/e 92 are observed (Figure 3).

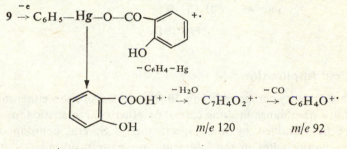

FIGURE 3

Methoxymethylmercuric chloride (5) and the corresponding ethoxy derivative (6) lose the alkoxy group to give a "base" peak at m/e 261. Both compounds exhibit strong M-1 ions in their spectra. It is noteworthy that in neither case a M-Cl peak is observed. Other fragmentations for 5 are shown in Figure 4 and major ions and their structure for 6 in Figure 5.

Compounds 7 and 8 both show, in addition to $Hg^{+\cdot}$, a characteristic ion at m/e 213 (CH_3Hg^+) as well as the respective molecular ions.

The sulfur-containing mercury compound 10 shows cleavage α to mercury, giving fragments at m/e 226 and m/e 153. The latter fragment in turn loses OH and CO.

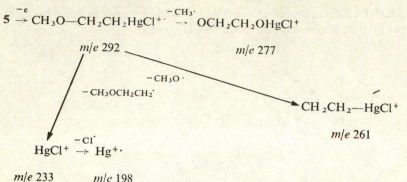

FIGURE 4

$$m/e\ 291 = M^{+\cdot}—CH_3{}^\cdot$$
$$m/e\ 261 = M^{+\cdot}—CH_3CH_2O^\cdot$$
$$m/e\ 233 = M^{+\cdot}-[CH_3CH_2O^\cdot+CH_2CH_2]$$
$$m/e\ 226 = M^{+\cdot}-[CH_3CH_2O^\cdot+Cl^\cdot]$$
$$m/e\ 198 = Hg^{+\cdot}$$
$$m/e\ \ 59 = CH_3CH_2O—CH_2{}^+$$
$$m/e\ \ 45 = CH_3CH_2O^+$$

FIGURE 5

Analytical Applications

All organomercury compounds **1** to **10** (Figures 6–10) show either molecular ions (of low abundance in some cases), or other characteristic ions and can therefore be identified by mass spectrometry. Several compounds show common intermediates in the fragmentation pattern, usually as abundant clusters of ions (e.g. $C_6H_5Hg^+$, $OCH_2CH_2HgCl^+$) and these as well as the ions for $Hg^{+\cdot}$ (cluster at m/e 198–204), which are present in all compounds, would indicate the presence of organomercury compounds even when observed in the spectra of crude extracts. A bar chart of abundances for mercury is otopesis shown in Figure 11 and the pattern of ions containing one mercury and one chlorine atom in Figure 12. Although the presence of mercury compounds of the type investigated can be recognized in crude extracts by medium-resolution mass spectrometry, high-resolution mass spectrometry with photographic recording is the superior method for identification of specific compounds under these conditions because of the low abundance of the molecular ion or characteristic ions in many instances. The photoplate can be exposed for prolonged periods of time and evaluation of the plates is aided by the mass deficiency of mercury isotopes. Photoplates

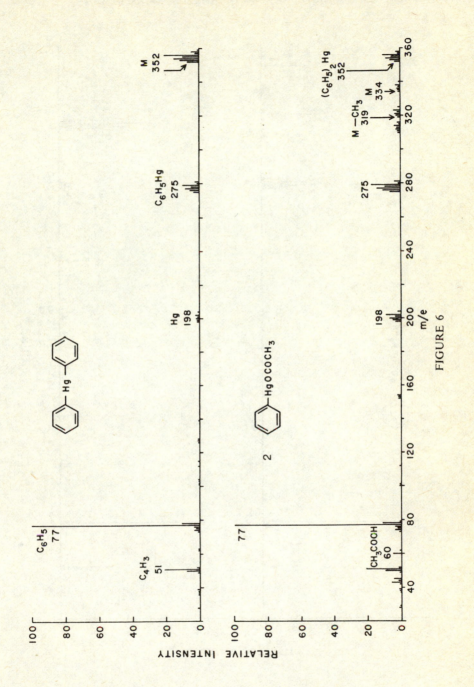

FIGURE 6

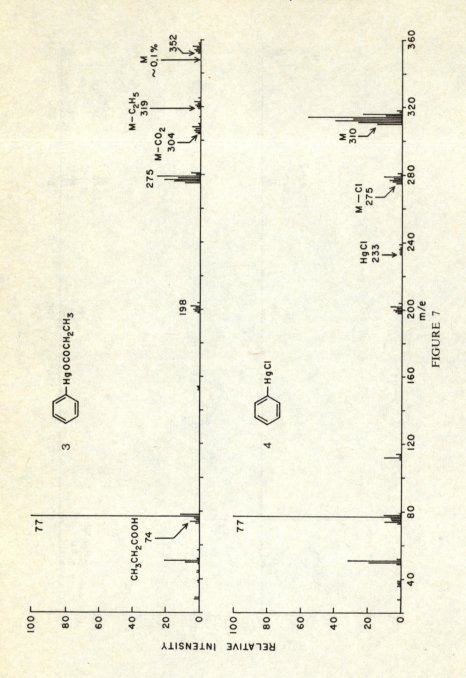

FIGURE 7

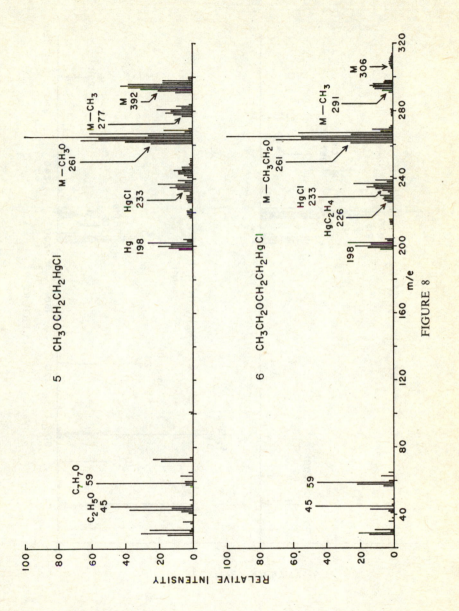

FIGURE 8

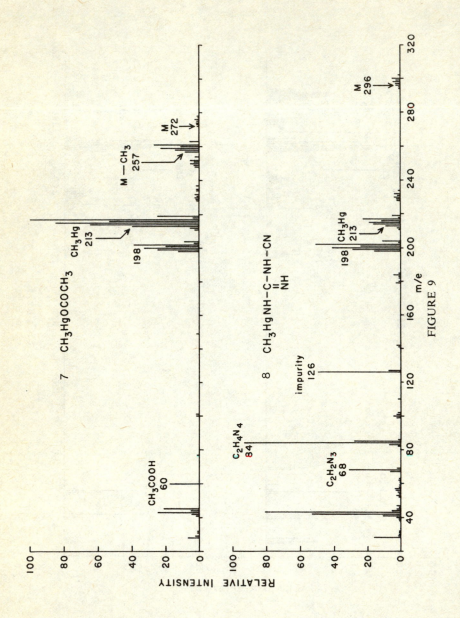

FIGURE 9

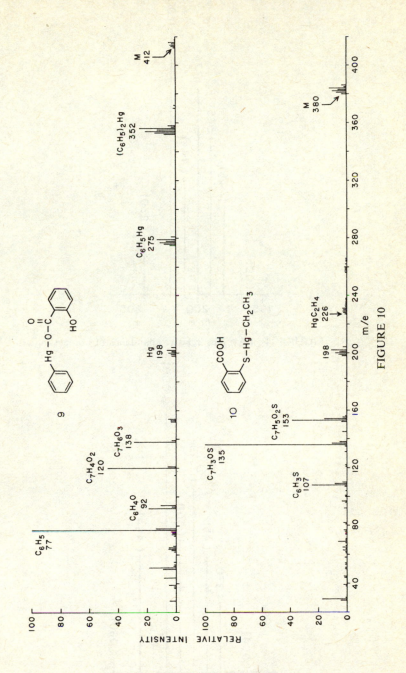

FIGURE 10

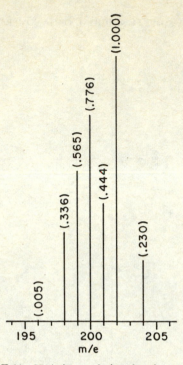

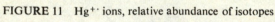

FIGURE 11 Hg$^{+ \cdot}$ ions, relative abundance of isotopes

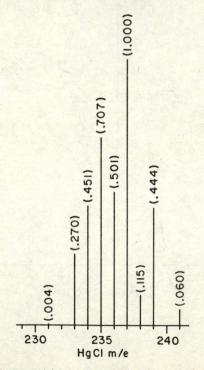

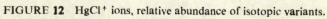

FIGURE 12 HgCl^{+} ions, relative abundance of isotopic variants.

exposed for 30 min to ions of samples A and B showed a line at m/e 296.04 (calculated for methylmercury dicyandiamide: 296.03), well separated from other ionic species of the same nominal mass.

Further application of high-resolution mass spectrometry to the identification of organomercury compounds in crude extracts, using automatic photoplate reading equipment, is in progress in these laboratories.

Acknowledgments

We thank D. J. Embree for recording the mass spectra and H. A. H. Wallace, R. D. Tinline and R. H. Burrage of the Canada Department of Agriculture for the samples of seeds treated with organomercury fungicides.

During the course of the preparation of this manuscript a study of electron-impact-induced fragmentations of organomercury compounds, including compounds **1** and **4**, has been reported.[26]

References

1. Swedish Royal Commission on Natural Resources (Eds.), *The Mercury Problem, Oikos Suppl.* 9 (1967).
2. M. W. Miller and G. G. Berg (Eds.), *Chemical Fallout* (Thomas, Springfield, Ill., 1969).
3. K. Östlund, *Acta Pharmacol. Toxicol.* **27**, Suppl. 1 (1969).
4. J. M. Wood, F. S. Kennedy, and C. G. Rosen, *Nature* **220**, 173 (1968).
5. S. Jensen and A. Jernelöv, *Nature* **223**, 753 (1969).
6. N. A. Smart, *Residue Rev.* **23**, 1 (1968).
7. U. Ulfvarson, in *Fungicides*, edited by D. C. Torgeson (Academic Press, New York, 1969), Vol. II. p. 303.
8. I. U. P. A. C. Information Bulletin No. 33, 199 and 205 (1969).
9. G. Widmark, *J. Ass. Offic. Anal. Chem* **53**, 1007 (1970).
10. J. C. Hansen, *Nord. Hyg. Tidskr.* **51**, 1 (1970).
11. G. G. Linstedt, *Analyst* **95**, 264 (1970).
12. R. W. April, and D. N. Hume, *Science* **170**, 849 (1970).
13. M. T. Jeffus, J. S. Elkins, and C. T. Kenner, *J. Ass. Offic. Anal. Chem.* **53**, 1172 (1970).
14. K. Östlund, *Nord. Hyg. Tidskr.* **82** (1969).
15. G. W. Johnson and C. Vickers, *Analyst* **95**, 356 (1970).
16. R. Takeshita, H. Akagi, M. Fujita, and Y. Sagagami, *J. Chromatogr.* **51**, 283 (1970).
17. G. Westöö in *Chemical Fallout*, edited by M. W. Miller and G. G. Berg (Thomas, Springfield, Ill., 1969), p. 75.
18. O. Hutzinger and W. D. Jamieson, *Bull. Environ. Contam. Toxicol.* **5**, 587 (1970).
19. V. H. Dibeler and F. L. Mohler, *J. Res. Natl. Bur. Stand.* **47**, 337 (1951).
20. B. G. Hobrock and R. W. Kiser, *J. Phys. Chem.* **66**, 155 (1962).
21. B. G. Gowenlock, R. M. Haynes, and J. R. Majer, *Trans. Faraday Soc.* **58**, 1905 (1962).
22. R. Spielmann and C. Delaunois, *Bull. Soc. Chim. Belges* **79**, 189 (1970).
23. S. C. Cohen and E. C. Tifft, *Chem. Commun.* 226 (1970).
24. J. L. Maynard, *J. Am. Chem. Soc.* **46**, 1510 (1924).
25. W. D. Jamieson and F. G. Mason, *Rev. Sci. Instrum.* **41**, 778 (1970).
26. W. F. Bryant and T. H. Kinstle, *J. Organometal. Chem.* **24**, 573 (1970).

Collection and Determination of Mercury in Air

JAN SCULLMAN and GUNNAR WIDMARK

Institute of Analytical Chemistry, University of Stockholm, Stockholm 50, Sweden

A method has been developed for the determination of low concentrations of mercury in air (nanograms/m³), i.e. in the range of the believed natural levels of mercury in the atmosphere (20 ng/m³). Mercury vapour has been collected from up to 200 l of air in glass tubes containing thin films of gold on sieved ceramic powder. In the laboratory the absorbed mercury was then released into a quartz-window cell by heating the tube in an oven at 500°C.

In this paper it is demonstrated that, by using extremely thin films of precipitated gold, quantitative recovery is obtained and memory effects, which result from the use of thicker films, are avoided.

INTRODUCTION

Spectrophotometric techniques have often been applied to the determination of small amounts of mercury,[1-8] and in two of these studies[4,8] atomic absorption spectrophotometers have been used. With conventional flames and burners, the instrument has a relatively low sensitivity for mercury, *viz.* 0.5 ppm in the prepared sample.[9] However, the sensitivity can be greatly improved by using the "cold cell" technique, in which the mercury vapour derived from the sample is introduced into an absorption cell equipped with quartz windows.

In the use of such absorption cells, enrichment methods have been used with the dual purpose of collecting the mercury and eliminating interfering substances. The ability of mercury to amalgamate with certain metals has often been utilized. For example, Brandenberger and Bader amalgamated mercury from solution on to a copper wire,[4] and Ulfvarson used gold foil to collect mercury vapour from the dithizone-mercury complex after it had been

69

decomposed by heat.[6] In a later publication, Lidmus and Ulfvarson describe the use of granular gold to collect mercury obtained from biological material.[7] However, to the best of our knowledge, it has not been previously demonstrated that the recovery of mercury is markedly influenced by the character of the surface of the amalgamating metal.

This paper describes an atomic absorption spectrophotometric method for the determination of mercury, which utilizes a long, narrow absorption cell. To collect mercury from air, we have used filters containing only a very thin film of gold on sieved ceramic powder. These filters, in glass tubing, have been used for up to 200 l of air, but the maximum volume of air which may be handled has not yet been determined. We have found that collecting devices composed of foils or grains of gold show a considerable memory effect which therefore tends to result in determinations of poor reproducibility. Our filter construction shows no memory effect, provided that the layer is sufficiently thin, thus allowing accurate determination of nanogram quantities of mercury.

The method was originally developed for the determination of low concentrations of mercury in air, preferably less than those found naturally, i.e. 20 ng/m^3.[10] However, the present method seems to be equally applicable to any system where mercury occurs as a vapour or where it can be converted conveniently into the vapour form. Present sensitivity is less than one nanogram per cubic metre of air.

EXPERIMENTAL

Instrumentation

The instrument used was a Jarrell-Ash atomic absorption spectrophotometer, model 82–537, with a Servogor 10 mV recorder. The optical bench was lengthened, thus allowing a long gas absorption cell to replace the flame. Light from the mercury hollow cathode source was collimated by an auxiliary lens of 10 cm focal length and another such lens focussed the light on to the entrance slit of the monochromator. A schematic diagram is shown in Figure 1.

The absorption cell consisted of a quartz tube (length 500 mm, i.d. 9 mm) on to which two quartz windows were fused. The cell was fitted with two side capillary tubes, one inlet and one outlet (i.d. 2 mm). To avoid interfering absorption on the walls of the absorption cell and the tubes, these were heated in an oven kept at 360°C. This temperature was originally selected after some experimentation and it has not been altered in subsequent work.

A collecting tube containing the mercury obtained, after air sampling, was fitted to the inlet tube with a ground joint. The mercury was released by heating with a demountable oven at 500°C, which was applied to the tube, and the vapour was transferred by a slow stream of nitrogen from the collecting tube into the quartz cell. The nitrogen, from a cylinder, entered the collecting tube through a hypodermic needle inserted through the rubber membrane.

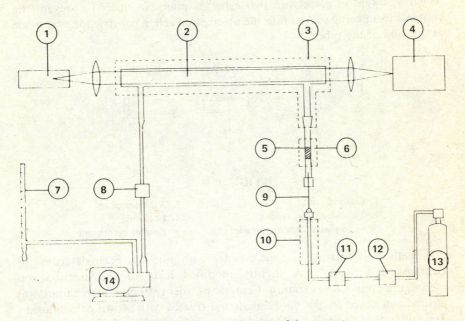

FIGURE 1   Schematic diagram of the apparatus.

1. Hollow cathode	2. Gas absorption cell
3. Heating oven	4. Monochromator
5. Collecting tube	6. Heating oven
7. Flow meter	8. Needle valve
9. Hypodermic needle	10. Heating oven
11. Valve	12. Gold trap
13. Nitrogen cylinder	14. Rotary pump

The nitrogen flow was regulated by a constant flow regulator and after passing through a gold trap[11] to remove mercury, was then preheated in an oven kept at 400°C. The absorption cell could be evacuated by a rotary pump connected to the outlet tube. This arrangement ensured constant low blanks during routine analysis.

The Collecting Tube

A sketch of the collecting tube is shown in Figure 2. The tube (i.d. 7 mm) was made of vycor glass and had a ground joint at one end. The other cooler end was fixed by means of epoxy resin on to a length of brass tubing with external threads. A rubber septum was attached to the brass tube by means of a compression nut. The overall length of the tube was about 150 mm.

The tube was filled to a length of 25 mm with gold-plated ceramic powder which was kept in place with two asbestos plugs. In order to prevent the contents from being sucked into the absorption cell, a constriction was made at one end of the tube.

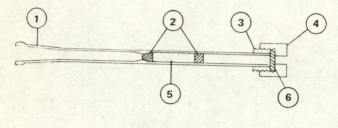

FIGURE 2

1. Ground joint
3. Threaded brass tubing
5. Gold plated ceramic powder
2. Asbestos plugs
4. Lock nut
6. Rubber membrane

A method for coating ceramic powder with gold has been described by Ormerod and Scott.[12] A slightly modified form of this method was used in the present investigation. Ceramic powder (340 mg), Sterchamol, was ground and sieved to pass 60–72 mesh and treated with brown gold chloride. Ten millilitre of a solution of the gold chloride in absolute alcohol, containing 15 mg of gold, was added to the ceramic powder and the mixture was degassed by rapidly applying vacuum from a water pump. After exactly 5 min under reduced pressure, when only part of the gold was reduced, the surplus liquid, together with some colloidal gold, was filtered off and the gilded powder was heated at 150°C for 2 hr. The actual amount of gold on the powder has not been determined, but it is assumed that the absence of a memory effect is due to the gold being thinly and evenly distributed on the ceramic surface.

ANALYTICAL PROCEDURE

On analysis, the monochromator is set so that maximum light transmission is obtained at the mercury resonance line of 253.7 nm. The carrier gas flow is then regulated to 55 ml of nitrogen per minute, a flow rate which has been found to give maximum response with the present instrumentation.

At the sampling location, the previously tested collecting tube is opened by removing both the nut and septum and the ground joint stopper. The latter end of the collecting tube is connected to a constant speed rotary pump through a flow meter. The maximum flow used was 700 ml/min at 25°C. After having been calibrated once, the gas volume passing through the tube is determined by the pumping time. During transport to the laboratory and during storage, the tube is closed tightly by the membrane and the glass stopper.

When a sample is to be analyzed at the laboratory, the collecting tube is attached to the absorption cell of the spectrophotometer and the carrier gas line is connected as described above. The demountable heating oven is then placed around the tube. After 1 min the mercury peak is recorded. Sometimes a preceding peak, originating from u.v.-absorbing organic materials is also found. This peak is, however, always to be found separated from the mercury peak. After sufficient training, one can carry out the whole measuring process at the spectrophotometer in about 2 min.

Standard samples and blanks were conveniently and frequently added to the collecting tube by injection through the rubber septum, using gas-tight syringes. This was most usually done when the collecting tube was attached to the gas cell, but some checks were made outside the laboratory. The standard gas samples were usually taken from 500-ml membrane-sealed glass flasks containing nitrogen and a droplet of mercury. Some empty flasks were also used to allow dilution of the standards. All flasks were kept in a box at a fairly constant temperature of $20 \pm 2°C$. It has been found that within 1 hr, five or more 0.5-ml aliquots can be withdrawn by syringe from the same flask, giving the same mercury value.

RESULTS AND DISCUSSION

Previous investigations of different kinds of collecting tubes, at this laboratory, have shown that the amount of gold in the amalgamating filters influenced the analytical results for small quantities of mercury. To study this pheno- menon in detail, four collection tubes were prepared which contained different amounts of gold as obtained by *total* reduction of the gold chloride added in alcoholic solution, thus following exactly the description of Ormerod and Scott.[12] Each tube contained 340 mg of inert carrier, Sterchamol, 60–72 mesh, and a gold load of 100, 10, 1, and 0.1 mg, respectively.

Typical absorption curves of such a study are given in Figure 3. It can be seen from this figure that with increasing amounts of gold and using the same volume of mercury-nitrogen gas, less mercury is released per unit time. Moreover, these collecting tubes show a time-dependent memory effect. Even tubes loaded with only 0.1 mg of gold showed slight memory effects, detected at injection from flasks. These slight losses were not easily detected, since the same recorded peak height was obtained as that given by the same amount of mercury vapour when injected into the absorption cell of the spectrophotometer (cf. Figure 3).

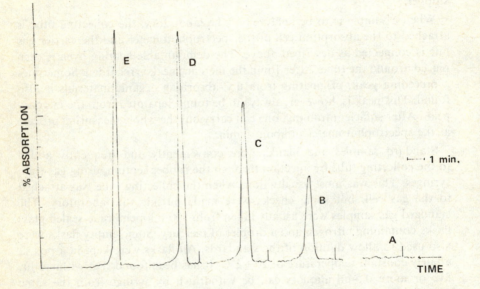

FIGURE 3   Recorder response for 10 ng of mercury released from collecting tubes plated with different amounts of gold. A, B, C, and D represent ceramic powder filters obtained from total reduction of gold chloride solutions containing 100, 10, 1, and 0.1 mg gold, respectively. E: 10 ng of mercury introduced directly into the absorption cell.

When applying small amounts of gold by precipitation according to the modified method described in this paper, the collecting tubes showed no detectable memory effects. This is probably due to the even distribution of the gold on the ceramic powder. A series of analyses was performed on four such collecting tubes. In this series, five known nanogram quantities of mercury in nitrogen were introduced into the tubes with a syringe. These samples were prepared by diluting some samples of nitrogen saturated with mercury. The results of these analyses are given in Table I and it is demonstrated that acceptable reproducibility is obtained.

Aerating mercury-loaded collection tubes with considerable volumes of air is not found to have any detectable effect on the size of the response obtained later on spectrophotometric analysis. However, this finding must not necessarily be taken as an indication that there are no losses when collecting large samples of air containing nanogram quantities of mercury. One of the most

TABLE I

Detection of mercury by the present method of atomic absorption

Hg (ng)	Collecting tubes†				Total‡	
	1	2	3	4		
2.7	0.0896	0.0998	0.0905	0.0941	0.0931	average absorbance
	0.0054	0.0070	0.0047	0.0032	0.0063	S.D.
	6.1	7.0	6.3	2.4	6.8	% rel. S.D.
4.5	0.1623	0.1663	0.1702	0.1559	0.1636	average absorbance
	0.0113	0.0106	0.0114	0.0030	0.0104	S.D.
	6.9	5.7	6.7	1.9	6.4	% rel. S.D.
6.9	0.3055	0.2878	0.3133	0.2860	0.2963	average absorbance
	0.0130	0.0127	0.0128	0.0057	0.0184	S.D.
	4.3	4.4	4.1	2.0	6.3	% rel. S.D.
8.8	0.4327	0.3989	0.4391	0.4015	0.4165	average absorbance
	0.0261	0.0251	0.0174	0.0107	0.0261	S.D.
	6.0	6.3	4.0	2.8	6.2	% rel. S.D.
9.9	0.5517	0.5286	0.5161	0.5260	0.5306	avarage absorbance
	0.0312	0.0341	0.0222	0.0287	0.0305	S.D.
	5.7	6.5	4.3	5.5	5.7	% rel. S.D.

†Each collecting tube has been used for five analyses of each amount of mercury in the range of nanograms as indicated.
‡Total refers to 20 analyses.

likely reasons for losses to occur is that, during the collection, reactive impurities of the air might oxidize the elementary mercury present in the sample. Although considered to be less probable, reduction might also occur in the collection tube when the air pumped through the tube contains some mercury salt aerosol. The possible presence of alkylmercury compounds in some air samples will contribute further to the complex nature of the analysis of mercury in air.

The results obtained with the present method of analysis on some air samples will be dealt with elsewhere.

References

1. O. Lindström, *Anal. Chem.* **31**, 461 (1959).
2. M. Schachter, *J. Ass. Offic. Anal. Chem.* **49**, 778 (1966).
3. A. R. Barringer, *Inst. Mining Met., Trans./Sect. B* **75**, 120 (1966).
4. H. Brandenberger and H. Bader, *Helv. Chim. Acta* **50**, 1409 (1967).
5. C. Ling, *Anal. Chem.* **39**, 798 (1967).
6. U. Ulfvarson, *Acta Chem. Scand.* **21**, 641 (1967).
7. V. Lidums and U. Ulfvarson, *Acta Chem. Scand.* **22**, 2150 (1968).
8. W. R. Hatch and W. L. Ott, *Anal. Chem.* **40**, 2085 (1968).
9. *Analytical Methods for Atomic Absorption Spectrophotometry, Perkin-Elmer* (Norwalk, Conn., U.S.A.).
10. E. Eriksson, The Mercury Problem, Symposium, Stockholm, 24–26 January, 1966. *Oikos Suppl.* (Copenhagen) **9**, 13 (1967).
11. J. E. Benson and G. S. Weiland, *J. Chem. Educ.* **41**, 223 (1964).
12. E. C. Ormerod and R. P. W. Scott, *J. Chromatogr.* **2**, 65 (1959).

Determination of Trace Levels of Mercury in Effluents and Wastewaters

K. H. NELSON, W. D. BROWN, and S. J. STARUCH

Burlington Industries Research Center, P.O. Box 21327, Greensboro, N.C. 27420, U.S.A.

Concern about mercury pollution of the environment and the inapplicability of natural water methods necessitated development of a procedure for determining parts per billion mercury in effluents and wastewaters containing large amounts of organic matter. The sample is digested with sulfuric and nitric acids to destroy the organic matter, and the ionic mercury is reduced to the elemental state by stannous ion. Then the digestate is aerated with a stream of air to carry the mercury vapor through a heated line into a quartz cell positioned in an atomic absorption spectrophotometer for measurement. Analyses of effluents and aqueous samples gave good recoveries of added mercury. Effluents, wastewaters, water supplies, and aqueous samples secured within manufacturing plants have been analyzed. With minor modification, the procedure has been applied to manufacturing materials such as vinylpyridine, latex, sizing, dyes, caustic, and hydro.

INTRODUCTION

The concern about mercury pollution in the environment, particularly in natural waters, has led to the development of analytical procedures suitable for determining this element in natural waters.[1-5] At the same time, this increasing concern has emphasized the need for rapid analytical capabilities to measure mercury at the parts-per-billion level in effluents discharged to the environment. Examination of the literature shows that, contrary to expectations, there appears to be little information available on determining mercury in wastewaters and effluents.

77

Most natural waters generally contain a relatively small amount of organic matter[6] which is easily destroyed through mild chemical oxidation. When determining mercury in natural waters, the water is acidified and then treated with potassium permanganate and potassium persulfate to destroy organic matter. After reduction to the elemental state, the resulting metallic mercury is usually collected by aeration for measurement, but in some procedures amalgamation and other techniques[1,2,7] are used to isolate the mercury. In the final step of the various procedures, the mercury vapor is measured by flameless atomic absorption.

In contrast to natural waters, many effluents from manufacturing processes often contain a variety of organic compounds[8] in amounts sufficient to interfere in many types of analyses. When the mercury methods for natural waters were applied to effluents from different sources, the results were not satisfactory. During the aeration of many treated effluents, copious amounts of foam filled the aeration apparatus and prevented measurement of the elemental mercury. Although addition of antifoam agents tended to suppress formation of foam, the agents had an adverse effect on the analytical data. Apparently the permanganate-persulfate reagent used in these methods does not completely destroy some types of organic matter present in effluents. Treatment of these effluents with other oxidants under similar conditions did not completely destroy all organic matter in many cases.

Because the existing methods proved unsuitable for effluents, a wet digestion procedure was developed to remove all organics and to eliminate the foaming problems. In this procedure, the sample is chemically treated to remove organic matter, decompose organomercurial compounds, and solubilize inorganic mercury compounds. The ionic mercury is then reduced to the elemental state by addition of stannous ion. Then the mercury vapour is carried by a stream of air or inert gas through a heated line into a quartz cell positioned in the light path of an atomic absorption spectrophotometer. The absorption of radiation at 253.7 nm, a function of the quantity of mercury present, is measured and recorded on a chart. The sensitivity and reproducibility of the method were increased by using a heated glass line rather than a desiccant in the aeration apparatus.

EXPERIMENTAL

Apparatus

Atomic absorption spectrophotometer   A Perkin-Elmer 303 atomic absorption spectrophotometer equipped with a recorder readout, a Perkin-Elmer 196 recorder, and a mercury hollow-cathode lamp were used. The burner head was replaced with a custom-made cell holder.

Aeration apparatus   The aeration apparatus, shown schematically in Figure 1, consisted of a cylinder of compressed air, a flowmeter, a three-way Teflon stopcock, a 250-ml gas washing bottle, a 10-cm quartz cell (Hellma No. 120-QS) with two filling tubes, and glass lines. The gas washing bottle, equipped with a medium-porosity fritted bubbler, had a 100-ml mark scribed on the outside. The quartz cell and the glass line from the gas washing bottle

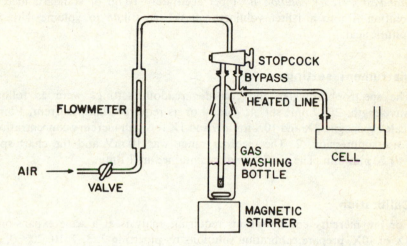

FIGURE 1 Aeration apparatus.

to the cell were heated by an electric heating tape. Viton O-ring ball and socket joints (Berkeley Glasslab Nos. BG 9017 and BG 9018) connected the stopcock to the heated line and the gas washing bottle. Teflon-coated ball and socket joints (Berkeley Glasslab Nos. BG 9028 and BG 9029) connected the heated line to the gas washing bottle and the quartz cell.

Reagents

Reagent-grade chemicals and distilled water were used for preparation of all solutions.

Sulfuric acid, 1 *N*   Add 83 ml of concentrated sulfuric acid to 3 l of distilled water in a glass bottle.

Sulfuric acid, 1 : 3   Add slowly, with stirring, 100 ml of concentrated sulfuric acid to 300 ml of distilled water. Cool and store in a glass bottle.

Stannous chloride solution   Dissolve 10 g of stannous chloride dihydrate in 85 ml of distilled water acidified with 15 ml of concentrated sulfuric acid.

Stock mercury solution   Dissolve 0.1000 g of redistilled mercury in a minimum amount of nitric acid. Transfer to a 1-liter volumetric flask and dilute to volume with 1 N sulfuric acid.

Standard mercury solution A   Pipet accurately 10 ml of the stock mercury solution into a 100-ml volumetric flask. Dilute to volume with 1 N sulfuric acid.

Standard mercury solution B   Pipet accurately 10 ml of standard mercury solution A into a 1-liter volumetric flask and dilute to volume with 1 N sulfuric acid.

Instrument settings

The spectrophotometer and recorder readout settings were as follows: wavelength, 253.7 nm; slit, 4; gain, 4 or as required; lamp current, 10 mA; scale expansion, 3X and 10X for low and 1X for high mercury concentrations; noise suppression, 2. The recorder range was 10 mV and the chart speed was 20 mm/min. The air flow was maintained at 1 l/min.

Calibration

For low mercury concentrations requiring analysis at a scale expansion of 3X or 10X, prepare calibration solutions by pipetting 2, 5, 7, 10, 20, 30, and 50 ml of standard mercury solution B into 1-liter volumetric flasks. Dilute each to volume with 1 N sulfuric acid. For the higher concentrations of mercury measured at a 1X setting, dilute 20, 40, 70, 100, and 130 ml aliquots of standard mercury solution B to 1 l with 1 N sulfuric acid.

Pipet 100 ml of a calibration solution into the gas washing bottle which contains the stirring bar. Inject 2 ml of the stannous chloride solution into the bottle and aerate as described in the section under "Procedure". Repeat twice for each calibration solution. Average the measured peak heights from three analyses of each calibration solution, and, if necessary, correct for the blank obtained with 1 N sulfuric acid. Plot calibration graphs of peak height versus mercury content in micrograms.

Procedure

Glassware preparation   Clean all glassware, cell, and glass lines with a dichromate-sulfuric acid solution, rinse thoroughly with distilled water, and dry in an oven at 105°C.

Sampling   Collect samples in jars, cleaned in the above manner, by rinsing the jar at least twice with the sample before filling and sealing.

Digestion To digest an effluent, shake the sample jar to thoroughly mix the sample and to determine if the sample foams. Then weigh duplicate 100-ml portions of the effluent into 125-ml Erlenmeyer flasks. Add 6 ml of concentrated sulfuric acid, 3 ml of concentrated nitric acid, and several porous silica boiling stones (Cargille) to each flask. If the sample foams when shaken, add three drops of Dow Corning Antifoam H-10 Emulsion (diluted 1:9) to each flask. Hang a glass hook on each flask lip, and place a small, short-stem funnel in each flask opening.

Place the flasks on a hot plate which is at a temperature of about 200°C. The flask contents will begin to boil smoothly in a few minutes without bumping or loss. Continue boiling until the sample is reduced to sulfuric acid fumes. When the digestate begins to darken, slowly flow down the flask walls from a dropper enough concentrated nitric acid to clarify the solution. Allow the nitrogen oxides to dissipate from the flask. Repeat this addition of concentrated nitric acid as needed until the digestate remains clear. Then flow two additional $\frac{1}{2}$-ml portions of concentrated nitric acid into the flasks in the same manner. Remove the flask from the hot plate and cool. With distilled water, rinse the glass hook and funnel and collect the rinsings in the flask. Then rinse down the flask walls. Bring the digestate to a boil for a few minutes to expel nitrogen oxides and then cool.

Aeration First, determine the blank and establish the cleanliness of the aeration apparatus. Add 10 ml of the 1:3 sulfuric acid to the gas washing bottle which contains the stirring bar. Bring the volume to 100 ml with distilled water. With the air flow diverted through the bypass and the recorder operating, syringe 2 ml of stannous chloride solution into the gas washing bottle. Immediately insert the bottle into the aeration apparatus and position the stirring motor. Allow the contents to mix for 1 min and then direct the air flow through the bottle. After approximately 3 min, divert the air flow through the bypass and remove the gas washing bottle from the apparatus. Rinse the bottle and the fritted bubbler several times with distilled water. Repeat the blank determination three or four times.

Next, aerate the digested samples. Use a stream of distilled water to quantitatively transfer a digestate into the gas washing bottle which contains 10 ml of 1:3 sulfuric acid and the stirring bar. Add sufficient distilled water to bring the volume to the 100-ml mark. Then treat the sample solution in the same manner as the blank determination. The mercury content of the sample will be recorded as a peak which is characterized by an immediate rapid rise and a decay to a baseline (Figure 2). After aeration, rinse the bottle as before and transfer the next digestate.

Calculation Measure the height of the peak perpendicular to a baseline drawn in the manner shown in Figure 2, curve B. If necessary, correct the peak height for the distilled water blank. Using the corrected peak height, read the micrograms of mercury in the sample from the appropriate calibration graph. Calculate the parts per billion of mercury and average the results of the duplicate determinations.

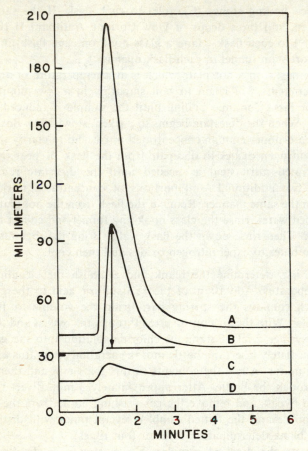

FIGURE 2   Measurement of peak height on typical peaks.

A = 0.5 mcg mercury at 3X
B = 0.07 mcg mercury at 10X
C = 0.006 mcg mercury at 10X
D = blank

DISCUSSION

The procedure has been routinely applied to several hundred samples of raw and treated industrial effluents. Although many of these samples contained unknown amounts and types of organic compounds from manufacturing operations, all were easily digested and analyzed for mercury. In addition, the procedure has been applied to municipal and industrial water supplies, and, with minor modification, to numerous proprietary chemicals.

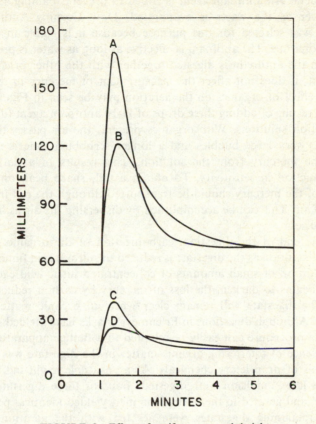

FIGURE 3   Effect of antifoam on peak height.

A = 0.33 mcg mercury
B = 0.33 mcg mercury with antifoam
C = 0.075 mcg mercury
D = 0.075 mcg mercury with antifoam

Some typical peaks obtained during analysis, together with the technique for measuring peak height, are shown in Figure 2. On the chart, the final baseline is extended to the front of the peak and the peak height is measured perpendicular to this line. At scale expansions of 3X and 10X, the peak height in millimeters is approximately proportional to absorbance and may be used as the ordinate on the calibration graph. For samples containing large amounts of mercury, the peak is recorded on the normal 1X setting. Then the peak height is measured in per cent absorption which is then converted to absorbance.

In the digestion, concentrated sulfuric acid and nitric acid are added to the effluent. An antifoam agent is added to prevent foaming and possible boiling over as the water is evaporated. Dow Corning Antifoam H-10 Emulsion was selected for this purpose because it is stable in hot, acidic aqueous solutions. The antifoam is effective as long as water is present in the flask; then the antifoam is digested together with the other organic matter. As a result, it does not affect the measurement of mercury in the aeration step. The effect of organics on the aeration may be seen in Figure 3, which shows the results of adding three drops of H-10 antifoam agent (diluted 1 : 9) to calibration solutions. With organics present, the air passes through the solution in very large bubbles and a longer scrubbing time is required to remove the mercury from the solution. This results in a flattened peak and a reduction in sensitivity. To obtain a tall, sharp peak on the chart, the bulk of the mercury should be transported through the cell in the initial volume of air. This can be accomplished by dispersing the air as fine bubbles in the solution.

Near the end of the digestion, carbonization of the organic matter will darken the solution as the digestate is reduced to sulfuric acid fumes. Through careful addition of small amounts of concentrated nitric acid each time the digestate begins to darken, the loss of mercury by carbon reduction is prevented. The digestate will remain clear when all organic matter has been destroyed. Although digestions in Erlenmeyer flasks have yielded satisfactory results, the procedure can easily be adapted to a Bethge apparatus.

The absence of interfering organic matter in the digestate was verified by the analysis of proprietary chemicals. As an example, eight individual portions of a latex emulsion were digested. Four of these digestates, selected at random and aerated in the normal manner, yielded identical peaks. When the four remaining digestates were aerated with the aeration apparatus modified by placing a $1\frac{1}{2}$-in. long tightly rolled silver screen plug in the heated line, no peaks were obtained on the recorder chart.

To ascertain that organomercurial compounds are decomposed and that the mercury is recovered, four compounds were digested and analyzed in triplicate. These compounds were phenylmercuric acetate, pyridylmercuric acetate, pyridylmercuric stearate, and di(phenylmercuric)dodecenylsuccinate. The first three were technical-grade products and the last compound, which contains 47.9% mercury, comprised 21% of Super Ad-It, a liquid fungicide additive for paints. From the stated mercury content, a calculated amount of a material was weighed into flasks so that each flask contained 100 mg of mercury. These samples were digested, appropriately diluted, and analyzed. Although the materials were of technical-grade quality, the data given in

Table I show that mercury can be recovered from organic compounds. The results for the three solid compounds are slightly high because of the non-homogeneity of these technical-grade products and a low stated assay. There was no evidence that any residual amounts of interfering organics, such as phenyl or pyridyl compounds, were present during measurement of the mercury.

TABLE I

Recovery of mercury from organomercurial compounds

	Mercury (mg)	
Compound	Taken	Found
Di(phenylmercuric)dodecenylsuccinate	100	101
	100	96
	100	105
Pyridylmercuric stearate	100	104
	100	120
	100	103
Phenylmercuric acetate	100	107
	100	116
	100	110
Pyridylmercuric acetate	100	112
	100	113
	100	121

The digestion procedure was applied to prepared samples containing 1, 5, and 10 ppb ionic mercury. These samples were prepared by dissolving 0.1000 g of metallic mercury in a minimum amount of nitric acid and then serially diluting with distilled water to obtain solutions with the final desired concentrations. The digestion procedure was carried out on 100-ml portions of the final solutions and the mercury was measured. The recoveries of mercury, given in Table II, show that ionic mercury is retained in the digestate.

Similarly, the recovery of mercury from five different effluents was studied. These effluents, which originated at different industrial locations, contained different amounts and types of soluble and particulate organic matter. Weighed 100-ml portions of each effluent were spiked by pipetting a known volume of an ionic mercury solution into each Erlenmeyer flask. To correct for any mercury that may have been present originally, duplicate 100-ml portions of each unspiked effluent were taken in separate flasks. All samples

TABLE II

Recovery of ionic mercury from solution

	Mercury (ppb)	
Sample	Taken	Found
A	1.0	1.1
B	1.0	1.0
C	1.0	1.4
D	5.0	5.0
E	5.0	5.0
F	5.0	5.0
G	10.0	10.0
H	10.0	10.0

TABLE III

Recovery of mercury from effluents

	Mercury (ppb)	
Effluent	Added	Recovered
A	1.0	1.1
	1.0	0.9
	1.0	1.0
B	1.0	1.0
	1.0	1.0
	1.0	1.1
	1.0	1.1
C	1.0	1.0
D	1.0	1.0
	1.0	0.9
E	5.0	4.9

were then digested and analyzed according to the procedure. The results, given in Table III, show that the presence of various unidentified organic compounds in the effluents has no effect on the recovery of mercury from the effluents.

Many proprietary chemicals in use today in industry may contain mercury. The mercury is present either as a residue from the preparative process or as an organomercurial compound added for fungicidal or bactericidal purposes. Various proprietary chemicals, such as latex, vinylpyridine, dye, stain, sizing, hydro, and caustic have been analyzed satisfactorily with the procedure (Table IV). Generally, these materials do not contain gross amounts of water,

TABLE IV

Analysis of proprietary chemicals

Chemical	Mercury (ppb)
Latex	15, 14
Latex	25, 28
Vinylpyridine	210, 220, 214
Vinylpyridine	7, 7
Rayon Dye	409, 410
Stain	184, 184
Sizing	0, 0
Hydro	0, 0
Caustic	24, 22
Caustic	19, 17

and the use of the antifoam agent can be omitted. To digest a proprietary chemical such as latex, triplicate 2-g portions are weighed into 125-ml Erlenmeyer flasks. Before placing the glass hook and funnel in the flask neck, 6-ml of concentrated sulfuric acid and 3-ml of concentrated nitric acid are added to each flask. With many samples, digestion begins immediately upon addition of acid and must be controlled by intermittent immersion of the flask in a cold water bath. Other samples may require a brief, gentle warming to initiate the digestion process. Incremental addition of concentrated nitric acid must be started immediately in order to minimize charring of organics and prevent loss of mercury. When the digestate remains clear, complete the digestion as with the effluents. Caustic samples do not require digestion unless the presence of organic substances is suspected. However, careful neutralization of the caustic with concentrated sulfuric acid is required before analysis or digestion.

References

1. G. W. Kalb, *At. Absorption Newslett.* **9**, 84 (1970).
2. M. J. Fishman, *Anal. Chem.* **42**, 1462 (1970).
3. S. H. Omang, *Anal. Chim. Acta* **53**, 415 (1970).
4. M. E. Hinkle and R. E. Learned, *U.S. Geol. Surv. Prof. Pap.* **650-D,** D251 (1969).
5. "Methods for Chemical Analysis of Water and Wastes 1971", Environmental Protection Agency, Water Quality Office, Analytical Quality Control Laboratory, Cincinnati, Ohio, p. 121.
6. K. H. Nelson and I. Lysyj, *Environ. Sci. Technol.* **2**, 61 (1968).
7. A. E. Moffitt, Jr. and R. E. Kupel, *At. Absorption Newslett.* **9**, 113 (1970).
8. J. J. Porter, D. W. Lyons, and W. F. Nolan, *Environ. Sci. Technol.* **6**, 36 (1972).

The Accurate Measurement of Lead in Airborne Particulates

A. ZDROJEWSKI, N. QUICKERT, L. DUBOIS, and J. L. MONKMAN

Air Pollution Control Directorate, Environmental Protection Service, Ottawa, Ontario, Canada

The accurate and rapid measurement of lead in airborne particulates is discussed with particular reference to so-called high-volume air samples. By standard addition analysis, it was found that there is no significant matrix error in measurements of lead taken on glass high-volume filters, within the range of concentrations investigated. This applies to varying or fixed amounts of the glass filter matrix. Recovery of standard added lead is quantitative. Although glass fibre sheet is much too impure for the analysis of most airborne metals, it is possible to analyze lead since the blank is reasonably uniform and since the amounts of lead on such a filter are large in comparison to the blank. Some additional corroboration has been obtained that the distribution of lead is uniform across the total exposed area of the filter. The final error of the lead measurement, as reported, is likely to be more largely a function of sampling problems such as time and flow measurements and particulate fall off and disturbance in sampling, transit and storage.

INTRODUCTION

Due to its toxicological importance, there is a voluminous literature on the measurement of lead. There is an increasing realization that the danger from inhalation may be greater than that from ingestion. Although the amount ingested in food and drink is much larger than that inhaled, the air absorption route may be more open and direct. Before going on to discuss atomic absorption (AA) as an analytical method, some of the other methods will be briefly mentioned.

(*a*) Polarography

(*b*) Anodic stripping voltammetry

(*c*) X-ray fluorescence

Polarographic methods for measurement of lead in air have been published by Levine[1] and by Dubois and Monkman.[2] The method is sensitive and accurate but the polarographic wave may be interfered with seriously if sample preparation is incomplete. Anodic stripping voltammetry is a very promising technique. It has been reviewed by Neeb[3] and some applications to the analysis of air samples have been given by Matson.[4]

X-ray fluorescence (XRF) has been suggested as a useful technique for the measurement of lead in airborne particulates which have been taken on dry filter media.[5] An advantage of XRF may be the minimal sample preparation where an aliquot disc of the filter is placed in the X-ray sample holder

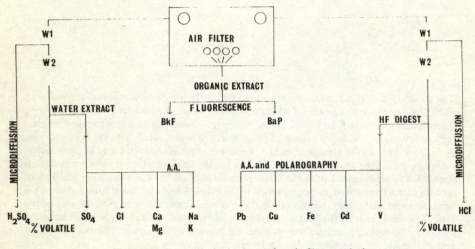

FIGURE 1   Analytical scheme for air filter analysis

without the need for a wet digestion procedure. A disadvantage of XRF is the inherent and appreciable matrix error. This error could be reduced if air samples were consistently taken on organic filter media. This is not the case at the present time, however, as high-volume air sampling networks in North America are set up on the basis of glass fibre filters. The matrix error due to the use of this very impure material is serious and not to be compensated for unless the standards can be made up in a background identical to the air samples. It is not sufficient, for example. to calibrate by comparing lead standards prepared in boric acid against airborne lead taken on a glass fibre filter. The calibration of an XRF method is likely to be tedious and somewhat difficult. In their work, Leroux and Mahmud[5] compared XRF measurements of lead against measurements already made by AA on the same samples, accepting the AA measurement as being correct.

Emission spectrography also suffers from matrix error and the associated need for lengthy calibration and carefully standardized conditions. Lead is not one of the elements best analyzed by emission spectrography. According to Morgan, AA is superior to emission spectrography.[6] While remembering that the emission spectrograph has been in use for a long time and is of classical respectability, a change in emphasis may now be noted, in that emission spectrography and XRF are now being used as scanning techniques for the preliminary assessment of a sample, by means of which a rough idea as to amounts and relative proportions of elements is obtained. After this, quantitative measurements are made by AA.

Figure 1 shows the procedure followed for the complete analysis of an air sample taken on glass filters. A description of the AA method used in our laboratory to measure lead in these samples will follow.

EXPERIMENTAL

Reagents

Air, compressed, in cylinders. Acetylene, compressed, in cylinders. Glass filters, 8×10 in., Gelman A. Filter paper, ashless, Whatman No. 41. Water, distilled at least twice from glass. Hydrofluoric acid, 49%, J. T. Baker Analyzed. Nitric acid, 71%, J. T. Baker Analyzed.

Apparatus

Atomic absorption spectrophotometer. Büchner funnel, polypropylene, $8\frac{1}{2} \times 10\frac{1}{2}$ in. Beakers, PTFE, 100-ml capacity.

PROCEDURE

Before using the glass filters for air sampling, the impurities present may be leached out by means of boiling distilled water. A special Büchner funnel, rectangular rather than circular, was constructed of polypropylene by Bel Art Products, Pequannock, New Jersey. This was designed to hold one box of 8×10 in. Gelman A filters. To improve the access of the hot distilled water, spacers (Teflon grids) are placed in the pile at intervals of every ten filters. The pile of filters is covered with boiling water, allowed to steep for 30 min and the water is withdrawn by suction. This process is repeated. The washed glass filters are allowed to air-dry in a dust-free location.

The glass filter is mounted in a conventional high-volume sampler. The air to be sampled is drawn through the filter at a flow rate of 40–50 cubic feet/min for 24 hr. The total volume of air sampled amounts to approximately 2000 m^3. Area aliquots are cut from the exposed area of the filter by means of a circular metal disc. One convenient area aliquot is a disc of 47-mm diameter. In preparation for analysis, one or more 47-mm discs are placed in a Teflon beaker. The glass filter matrix is dissolved by the drop-wise addition of 1 ml of hydrofluoric acid. The contents of the beaker are gently fumed at low heat until the hydrofluoric acid is evaporated. At this point, 1–2 ml of nitric acid

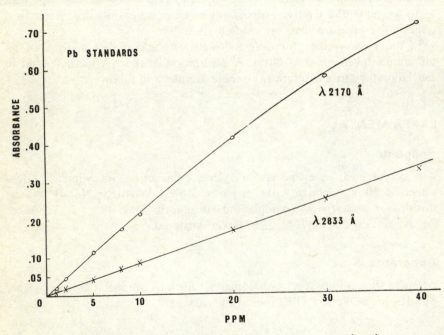

FIGURE 2   Aqueous standard lead curves at two wavelengths.

is added and gentle heating is continued until a few drops of HNO_3 are left. About 10 ml of distilled water is added, brought close to the boiling point and the sample is filtered through Whatman No. 41. The beaker is rinsed down with 10 ml distilled water, which is similarly heated and filtered. The combined filtrates are transferred to a 25-ml volumetric flask and made up to volume and also stored in screw-cap polyethylene bottles. The sample is now ready for measurement.

Figure 2 shows the calibration curves obtained at 2833 Å and 2170 Å with a commercially available AA instrument, PE403, using standards diluted with distilled water. Curve 1 of Figure 3 is the calibration curve obtained again at 2833 Å with standards diluted in the filter blank. This filter blank was obtained by digestion with HF and HNO_3 of a 47-mm disc, as described previously, and subsequent dilution to 25 ml with distilled water. Since this second curve intersects the Y axis, this intercept value can be considered as the amount of lead present in the filter, or as the amount of lead and other interfering materials giving a response as lead.

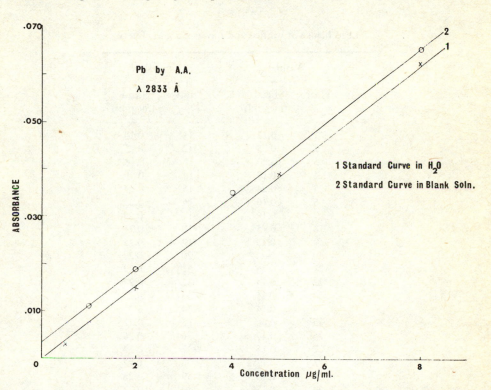

FIGURE 3 Standard lead curves in water, and filter matrix.

Repetitive analysis of 40 different filters, 20 washed filters and 20 unwashed, gave respectively an average value of 0.19 mcg/ml and 0.24 mcg/ml for a total of 4.75 mcg and 6.0 mcg for a 47-mm disc. The values are given in Table I. It should be noted, however, that these values are related to a given batch of filters only and values as high as 0.40 mcg were observed with a different batch of filters. Results obtained in the following experiments were calculated from the calibration curve prepared with the standards diluted in distilled water.

One aspect of the practical reproducibility obtained by this method is illustrated by Table II which shows the satisfactory agreement obtained for a group of ten replicates cut from one particular high-volume filter. The agreement is so good that in most cases differences are only evident in the second decimal of micrograms. The uniformity of these values supports the theory that the distribution of lead across the filter is also uniform.

TABLE I

Lead blanks of washed and unwashed glass filters

	Washed		Unwashed	
	No.	Lead (mcg/ml)	No.	Lead (mcg/ml)
	1	0.13	11	0.14
	2	0.28	12	0.25
	3	0.20	13	0.18
	4	0.18	14	0.19
	5	0.24	15	0.27
	6	0.16	16	0.23
	7	0.19	17	0.29
	8	0.15	18	0.21
	9	0.23	19	0.25
	10	0.24	20	0.20
	21	0.26	31	0.27
	22	0.24	32	0.27
	23	0.25	33	0.25
	24	0.14	34	0.25
	25	0.06	35	0.22
	26	0.17	36	0.26
	27	0.15	37	0.22
	28	0.12	38	0.27
	29	0.18	39	0.28
	30	0.17	40	0.22
n	20	- -	20	- -
Mean	- -	0.19	- -	0.24
Low	- -	0.13	- -	0.14
High	- -	0.28	- -	0.29

In the analysis of air samples taken on an 8×10 in. glass fibre filter, it is taken for granted that the distribution of the contaminant is uniform across the whole exposed area. This is implicit in the methods of sample preparation which are usually based on the preparation of a portion of the filter. The calculation of the total quantity of contaminant is made by prorating the aliquot area and total exposed area. This seems a dangerous assumption to make; however, experiments carried out by Dubois *et al.*[7] indicated that there was no significant difference in the lead levels obtained by analyzing aliquot discs taken from different portions of the filter. Since only one aliquot size was considered in this work, it may be that significant differences would begin to appear if the aliquots were sufficiently reduced in size. It is also necessary to remember that, if the high-volume filter is overloaded, classification may occur and particulates may be lost by fall off. The latter situation will result

TABLE II

Ten replicate lead analyses from one filter

First aliquot	First reading (mcg/ml)	Second reading (mcg/ml)	Difference
A	1.65	1.67	0.02
B	1.78	1.87	0.09
C	1.73	1.85	0.12
D	1.67	1.64	0.03
E	1.75	1.81	0.06
F	1.47	1.58	0.11
G	1.67	1.69	0.02
H	1.62	1.61	0.01
I	1.73	1.78	0.05
J	1.61	1.73	0.12
Arithmetic mean	1.67	1.72	0.06
S.D.	0.085	0.096	0.04
Coefficient of variation	5.08	5.56	--

in analytical results lower than the truth. This possibility is probably best prevented by standardizing sampling conditions at a lower flow rate or for a shorter time period.

Taking two instrumental readings on the same prepared sample, as in Table II, gives an indication of the reproducibility of the AA instrument quite independently of the chemical treatment of the sample. A typical example of the reproducibility of the method is shown in Table III, where 62 air samples were analyzed in duplicate. Two separate 47-mm discs were cut from each filter, separately digested and the duplicate prepared samples were independently assayed for lead by AA. The arithmetic means and standard deviations of the differences between measurements are given in micrograms per ml of prepared sample. The difference in the average lead concentrations

TABLE III

Duplicate lead analyses of 62 different air filters

Sample No.	First measurement (mcg/ml)	Second measurement (mcg/ml)	Difference
1	1.64	1.70	0.06
2	1.57	1.53	0.04
3	2.76	2.88	0.12
4	11.08	10.74	0.34
5	5.15	5.28	0.13
6	5.46	5.44	0.02
7	8.30	8.24	0.06
8	7.65	7.50	0.15
9	1.96	2.01	0.05
10	2.71	2.86	0.15
11	1.61	1.65	0.04
12	0.72	0.79	0.07
13	2.01	1.53	0.48
14	1.85	1.79	0.06
15	1.68	1.73	0.05
16	1.27	1.26	0.01
17	2.73	2.69	0.04
18	4.66	4.66	0.00
19	4.06	3.99	0.07
20	2.08	2.14	0.06
21	1.45	1.43	0.02
22	1.33	1.30	0.03
23	2.02	2.08	0.06
24	2.78	2.58	0.20
25	3.25	3.16	0.09
26	2.15	2.07	0.08
27	5.22	5.37	0.15
28	2.11	2.02	0.09
29	2.73	2.72	0.01
30	1.08	1.06	0.02
31	2.65	2.62	0.03
32	6.14	5.92	0.22
33	1.98	1.86	0.12
34	1.72	1.72	0.00
35	4.22	4.08	0.14
36	1.96	1.98	0.02
37	1.73	1.83	0.10
38	8.29	8.14	0.15
39	1.72	1.71	0.01
40	1.63	1.85	0.22

TABLE III—*cont.*

Sample No.	First measurement (mcg/ml)	Second measurement (mcg/ml)	Difference
41	3.96	3.84	0.12
42	1.71	1.76	0.05
43	3.81	3.85	0.04
44	6.50	6.52	0.02
45	0.87	0.88	0.01
46	1.82	1.94	0.12
47	1.42	1.25	0.17
48	1.22	1.24	0.02
49	2.13	2.17	0.04
50	2.17	2.19	0.02
51	0.82	0.95	0.13
52	2.19	2.39	0.20
53	1.50	2.01	0.51
54	1.74	1.91	0.17
55	4.44	4.45	0.01
56	4.91	4.96	0.05
57	1.37	1.53	0.16
58	5.39	5.70	0.31
59	5.61	5.02	0.59
60	1.19	1.15	0.04
61	2.60	2.66	0.06
62	2.04	2.02	0.02
Arithmetic mean	2.99	3.00	0.11
S.D.	2.12	2.07	0.11

of the two separate groups of 62 assays is 0.11 mcg/ml on a mean value of 3.00 mcg/ml.

Based upon the statistical analysis of the lead values found in 38 air samples Burnham *et al.*[8] concluded that "it is necessary to utilize the standard additions technique to overcome matrix effects". Experiments carried out by us on blank glass filters, on air samples taken on such filters and on aqueous lead standards, are summarized in Figure 4 which does not support this statement. Figure 4 represents, firstly, lead standards in water run at 2833 Å; secondly, lead standards added to an air sample having a particulate loading of 36 mcg/m^3, and, thirdly, lead standards added to an air sample having a particulate loading of 143 mcg/m^3.

If a matrix effect exists, the effect can be either negative or positive, resulting in a decrease or increase in the measured lead value as compared with the truth. The curves of Figure 4 show that the slopes obtained with standard addition of lead are identical to the slope of the calibration curve obtained with lead standards made up in distilled water. In Figure 5 the same samples and standards are also analyzed at 2170 Å. The results obtained at 2170 Å are in complete agreement with those obtained at 2833 Å. An additional proof of no interference from the matrix can be derived from the results given in Table IV and V. In Table IV six separate identical volume aliquots were taken from prepared air sample No. 758. The volume aliquots were made up to

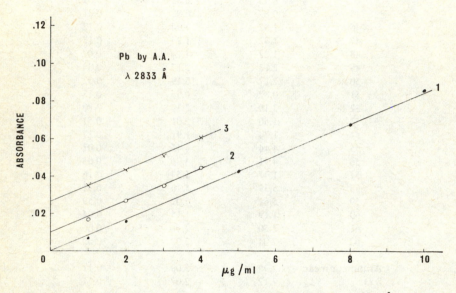

FIGURE 4   Lead standards in three different matrices at 2833 Å.

differing total volumes. The values for total lead found are identical for each of the six assays, within experimental error. Similarly, in Table V, seven separate but differing volume aliquots were taken from prepared air sample No. 759. One hundred micrograms of standard lead were added to each of the aliquots, which were made up to total final volumes varying from 5.0 to 50.0 ml. Again, all lead assays are within experimental error, with the exception of one. In Table VI, ten replicate discs were taken from one air filter, these were prepared for analysis with the standard addition of 50.0 mcg of lead in each case. Three blank discs from three separate filters were prepared for analysis at the same time. The results of Table VI indicate not only the lack of matrix effect but the uniformity of distribution of lead particulates across the surface of the filter.

Burnham *et al.*,[8] working with air samples taken on glass fibre filters, also suggest that lead losses could occur due to the treatment of the sample. In their sample preparation, they make use of a dry ashing step at 500°C for

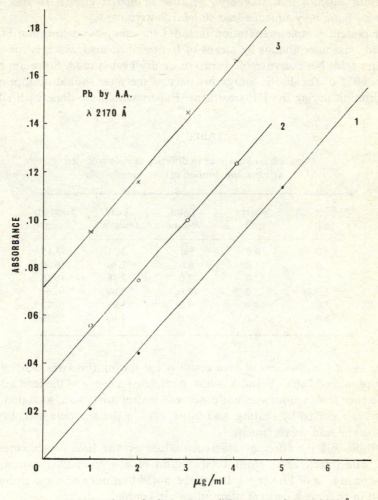

FIGURE 5 Lead standards in three different matrices at 2170 Å.

1 hr. Indeed, it is conceivable that lead may be lost by volatilization at this temperature, made unavailable by the formation of refractory oxides insoluble in nitric acid, or made unavailable by reaction with the glass fibre to form a lead glass. It is at least partly for reasons of this kind that we employ a wet ashing procedure at low temperatures.

Hwang and Feldman[9] worked with air samples taken on organic membranes. Their results, for this reason, will be free of the "matrix" effects of glass fibre. According to them it is not possible to measure airborne lead by AA using aqueous lead standards because of matrix effects. In the work presented here, only aqueous lead standards were used.

Our present sample preparation includes the complete dissolution of the exposed glass filter aliquot by means of hydrofluoric acid which is followed by nitric acid. No excessive temperature or dry heat is used. Since approximately 60% of the disc is SiO_2, this part of the filter should disappear by volatilization during the HF treatment. Experiments were done with added

TABLE IV

Replicate lead analyses in differing sample volumes
(matrix and dilution errors, sample 758)

Aliquot 758 (ml)	Water (ml)	Total volume	Lead (mcg/ml)	Total Pb (mcg)
5.0	0.0	5.0	3.31	16.55
5.0	1.0	6.0	2.80	16.80
5.0	2.0	7.0	2.38	16.66
5.0	3.0	8.0	2.08	16.64
5.0	4.0	9.0	1.88	16.92
5.0	5.0	10.0	1.68	16.80

lead to see if possible loss of lead could occur during this treatment of the filter. Results of Tables V and VI show complete recovery of the lead added. Interference from anions was not observed; hydrofluoric acid, as stated previously, is removed by heating, and nitric acid in the amounts used has no effect on the lead measurement.

In Table VII the effect of different cations on the lead measurement is listed. The interferences from 1000 mcg/ml of a given added cation are expressed in mcg of Pb per ml. It is to be noted that none of these cations is present in so large a concentration in an air sample.

The precision of lead determinations as carried out by Burnham et al.[8] and in this laboratory is compared in Table VIII. In both cases the Pb absorption at 2170 Å was used. The results of Burnham et al. were obtained with a Perkin-Elmer 303 spectrophotometer, whereas the results from this laboratory were obtained using the Perkin-Elmer 403 spectrophotometer with digital readout. The table shows that our coefficient of variation is still less than 2% for a 1-ppm solution using the 100 average mode. This is a

TABLE V

Replicate lead analyses with constant added lead and varying matrix†

Aliquot 759 (ml)	Total volume	Lead found (mcg/ml)	Total Lead	Less standard	Blank lead (mcg/ml)
4.0	5.0	26.80	134.00	34.00	8.50
9.0	10.0	18.25	182.50	82.50	9.17
14.0	15.0	15.30	229.50	129.50	9.25
19.0	20.0	14.00	280.00	180.00	9.47
24.0	25.0	12.91	322.75	222.75	9.28
49.0	50.0	11.15	557.50	457.50	9.34
50.0	50.0	9.30	465.00	465.00	9.30

†100 mcg lead standard added to each aliquot.

TABLE VI

Ten replicate lead analyses with constant added lead and constant matrix

Sample No.	First reading (mcg/ml)	Second reading (mcg/ml)	
1	2.36	2.34	
2	2.45	2.43	
3	2.33	2.31	
4	2.42	2.43	
5	2.29	2.27	
6	2.41	2.50	
7	2.30	2.33	
8	2.37	2.42	
9	2.24	2.32	
10	2.41	2.49	Total mean = 2.37
11 Blank	0.31	0.42	
12 Blank	0.31	0.39	
13 Blank	0.40	0.40	Total mean = 0.37

By subtraction 2.00 mcg/ml
Lead found $2.00 \times 25 = 50.00$ mcg
Lead added = 50.0 mcg

TABLE VII

Ion	Interference as Pb (mcg ml)	Concentration in extract (mcg/ml)	Estimated interference
Na	0.04	400	0.02
Al	0.15	200	0.03
Si	0.08	400	0.03
Ba	0.00	300	0.00
Zn	0.03	150	0.00
Ca	0.10	200	0.02
K	0.00	150	0.00
Mg	0.00	10	0.00

†Interfering ions assayed at concentration of 1000 mcg/ml.

TABLE VIII

Precision of lead determinations

Conc. (ppm)		Average absorbance	S.D.	% Coeff. of variation	No. of analyses
H_2O	A†	0.0093	0.0020	21.5	9
Blank	B	0.0001	0.0003	—	10
	C	0.0003	0.0009	—	10
1	A	0.0176	0.0034	19.3	9
	B	0.0292	0.0005	1.7	10
	C	0.0287	0.0011	3.8	10
2	A	0.0362	0.0017	4.7	5
	B	0.0567	0.0006	1.1	10
	C	0.0553	0.0012	2.2	10

†A   C.D. Burnham et al.
 B   This lab, using the 100 average mode on digital readout.
 C   This lab, using the 10 average mode on digital readout.

considerable improvement over the results of Burnham et al., which already show a 19.3% variation for this concentration. Comparison of measurements B and C indicates that the standard deviation decreases by about a factor of 2 when the total number of readings is increased by a factor of 10. This is in approximate agreement with statistics, which predict a decrease by a factor of $\sqrt{10} = 3.2$.

DISCUSSION

Our method of measuring lead in airborne particulates taken on glass fibre sheet is accurate and free of interferences in the range of 0–20 mcg of lead per ml of prepared solution, the average difference between measurements being 0.11 mcg/ml. This is true for areal sample sizes from 4–8% of the exposed area of an 8×10 in. filter. The overall accuracy of the measurement of lead in air is the sum of individual accuracies, or otherwise expressed, the overall error is the sum of the individual errors. These sources of error include the analytical error and the error in measuring the volume of air sampled.

Lead levels in air are usually expressed in micrograms per cubic metre of air sampled. To get this number, it is necessary to divide the analytical result (in micrograms) by the volume (in cubic metres). The errors in the volume are made up of the error in measuring the flow rate and the error in the time measurement. Although no experimental results have been obtained on the experimental errors associated with sampling, it is appropriate to say that the chemical measurements, as described previously, contribute a minimum to the overall error of the lead measurement in air samples.

References

1. L. Levine, *J. Ind. Hyg. Toxicol.* **27**, 171 (1945).
2. L. Dubois and J. L. Monkman, *Amer. Ind. Hyg. Assoc. J.* **25**, 485 (1964).
3. R. Neeb, *Angew. Chem. Intern. Ed.* **1**, 196 (1962).
4. W. R. Matson, Proceedings of the University of Missouri's 4th Annual Conference on Trace Substances in Environmental Health, June (1970).
5. J. Leroux and M. Mahmud, *J. Air Pollut. Control Assoc.* **20**, 402 (1970).
6. G. B. Morgan, Presented at the Pittsburgh Conference on Analytical Chemistry and Applied Spectroscopy, Pittsburgh, March (1967).
7. L. Dubois, T. Teichman, J. M. Airth, and J. L. Monkman, *J. Air Pollut. Control Assoc.* **16**, 77 (1966).
8. C. D. Burnham, C. E. Moore, and E. Kanabrocki, *Environ. Sci. Technol.* **3**, 472 (1969).
9. J. Y. Hwang and F. J. Feldman, *Appl. Spectrosc.* **24**, 371 (1970).

A Radiochemical Method for Selective Determination of Traces of Lead

J. STARÝ and K. KRATZER

Department of Nuclear Chemistry, Technical University of Prague, Břehová 7, Praha 1, Czechoslovakia

A very simple and rapid radiochemical method for the determination of 0.01—1 mcg of lead has been developed. It consists of adding carrier-free ^{212}Pb to the analysed sample, followed by shaking with a standard lead diethyldithiocarbamate solution in carbon tetrachloride. The amount of non-active lead in the sample is determined from activities of the organic and aqueous phases. A 10-fold excess of bismuth and more than 50-100-fold excess of other metals extractable as diethyldithiocarbamates do not interfere in the determination.

INTRODUCTION

Traces of lead are usually determined by solvent extraction with dithizone, using potassium cyanide as masking agent;[1,2] however, thallium(I) and bismuth in equal amounts and iron(III) in greater excess seriously interfere and have to be removed. Moreover, the dithizone method, as well as other analytical methods, requires the quantitative isolation of lead from the sample, which is in many cases rather difficult. This difficulty can be avoided by using carrier-free ^{212}Pb, as the separation yield can be determined easily from the radioactivity measurements.

The aim of the present paper was to develop a simple, rapid, selective and sensitive radioanalytical method for the determination of traces of lead.

105

Neutron-activation analysis is not suitable for this purpose because of its low sensitivity. Substoichiometric determination of lead with dithizone is selective but not very sensitive[3] due to the low value of the extraction constant K_{ex} (log K_{ex} = 1.0 for carbon tetrachloride as organic solvent).[4] The extraction of lead diethyldithiocarbamate into carbon tetrachloride is much more suitable in this respect because of the high stability of this chelate (log K_{ex} = 8.0).[4] Unfortunately, our preliminary experiments showed that the extraction of lead with a substoichiometric amount of zinc diethyldithiocarbamate (this reagent is more stable than diethyldithiocarbamic acid itself) is rather slow. During these experiments it was found that the two-phase isotope exchange between lead diethyldithiocarbamate in the organic phase and lead tartrate in the aqueous phase is rapid even for very low concentrations of lead and this fact has been utilized in the development of the present method. It is of interest to note that the two-phase isotope exchange between bismuth diethyldithiocarbamate and bismuth tartrate complex in similar conditions is extremely slow.[5]

THEORETICAL

The principle of the present method consists of adding carrier-free ^{212}Pb to the analysed sample containing an unknown amount (x) of lead. Then ammonium tartrate is added and the solution is shaken with standard lead diethyldithiocarbamate in carbon tetrachloride containing a known amount (m) of lead. After the isotope equilibrium is reached the specific activities of lead are distributed in the aqueous and organic phases according to the following equation:

$$\frac{A_{aq}}{x} = \frac{A_{org}}{m} \tag{1}$$

A_{aq} and A_{org} denote the measured activities of lead in the aqueous and organic phases, respectively.

The practical application of the two-phase isotope exchange in trace analysis requires a high stability of the chelate in the organic phase in the absence of the excess of organic reagent even at highest dilutions. The optimum conditions for the determination can be predicted from the theory of substoichiometry:[6]

$$pH > -0.01 \log c_{PbA_2} - \tfrac{1}{2}\log K_{ex} + \tfrac{1}{2}\log (1 + \beta_1[Tart^{2-}]) \tag{2}$$

where c_{PbA_2} is the concentration of lead diethyldithiocarbamate, β_1 is the stability constant of lead tartrate (log β_1 = 2.9)[2], and [$Tart^{2-}$] is the equilibrium concentration of tartrate anion. For the determination of submicrogram amounts of lead the value of c_{PbA_2} has to be of the order of 10^{-6} —

10^{-7} M. From Eq. (2) it is evident that pH has to be higher than 6 when $[Tart^{2-}] = 0.2$ M.

The extraction of lead as diethyldithiocarbamate in the presence of potassium cyanide is rather selective; only thallium(I), thallium(III), bismuth and partially also indium and antimony(III) are extracted simultaneously with lead.[2] In the two-phase isotope exchange reaction, thallium(I), indium and antimony(III) will not interfere, because their extraction constants are much lower than those of lead.[2,4] The interference of bismuth and thallium(III) can be substantially decreased by the presence of tartrate which forms much more stable complexes with these metals than with lead.[2]

EXPERIMENTAL

Apparatus and equipment

A pH-meter (Radiometer TTT-1, Copenhagen) with glass electrode and a scintillation counter with the well-type NaI(Tl) crystal (channel 0.238 MeV) were used. Reactions were carried out in 20-ml glass test tubes with ground-glass stoppers.

Reagents

Unless otherwise stated, all reagents, including sodium diethyldithiocarbamate, were of analytical-reagent-grade purity. Carbon tetrachloride was distilled twice. The preparation of the buffer solutions was as follows. Buffer solution A. Dissolve 11.5 g of sodium tartrate and 3.3 g of potassium cyanide in 50 ml of water. Adjust pH with sodium hydroxide to about 12. Buffer solution B. Dissolve 9.2 g of ammonium tartrate and 3.3 g of potassium cyanide in 50 ml of water (pH about 9.5).

To prepare the standard lead diethyldithiocarbamate solution add to 20 ml of aqueous lead nitrate solution (pH 2–6), containing 20.0 or 2.0 mcg of lead, 5 ml of buffer solution A, 60 ml of carbon tetrachloride, and 1–2 mg of solid sodium diethyldithiocarbamate. Shake for 2 min and separate carefully the organic extract containing 1.0 or 0.1 mcg Pb/3 ml. Thus prepared, the solution contains less than 10^{-9} M free diethyldithiocarbamic acid. Carrier-free ^{212}Pb is prepared from RdTh (^{228}Th, T = 1.91 y) solution (pH ~ 2) which is in the radioactive equilibrium with the daughter nuclides ^{224}Ra (T = 3.64 day), ^{220}Rn (T = 51.5 sec), ^{216}Po (0.158 sec), ^{212}Pb (10.5 hr), ^{212}Bi (60.5 min), ^{208}Tl (3.1 min) and ^{212}Po (3×10^{-6} sec). To 2.5 ml of the RdTh solution in a glass test tube 0.5 ml of buffer A, 6 ml of carbon tetrachloride and about 0.1 mg of solid sodium diethyldithiocarbamate are added. Shake for 2 min, transfer the organic extract into another test tube and reextract ^{212}Pb into 2 ml of 1 M hydrochloric acid (2 min shaking). Separate the aqueous phase and dilute 20 times with water (activity about 10^5 cpm/ml).

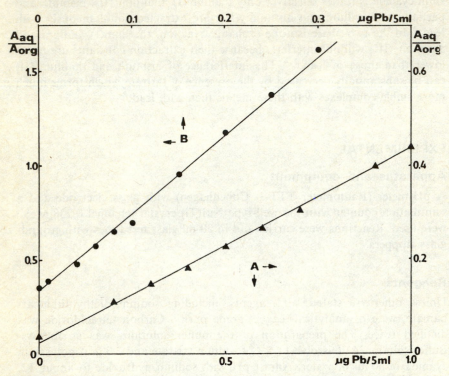

FIGURE 1   Calibration curve for the determination of lead. The amount of lead in standard lead diethyldithiocarbamate solution is 1.0 mcg (Curve A) and 0.1 mcg (Curve B).

Two-phase isotope exchange

The time necessary for attaining complete isotopic exchange between lead tartrate and lead diethyldithiocarbamate was studied radiometrically, labelling alternately the first or the second complex with ^{212}Pb. It was found that at pH 9.5 the equilibrium was reached in 1–2 min in the concentration range studied. In the absence of tartrate or at lower pH-value the kinetics of the isotopic exchange is substantially slower.

Procedure

To 5.0 ml of an analysed sample (pH 2–6), labelled with carrier-free ^{212}Pb, add 1.0 ml of buffer B (final pH∼9.5) and shake for 2–4 min with 3.0 ml of standard lead diethyldithiocarbamate solution in carbon tetrachloride. Measure the activity of 2.0-ml aliquots of the organic and aqueous phases immediately (channel 0.238 MeV) or after 5 hr (after reaching the radioactive equilibrium between ^{212}Pb and ^{212}Bi).

Influence of foreign ions

The possible interference of a great number of metals in the determination of lead by the above method was studied in the following way: samples containing 1.00 mcg of lead and 10–100 mcg of other metals were analysed for the lead content as described in Procedure.

TABLE I

Determination of lead in the presence of foreign metals

Lead present (mcg)	Foreign metals present (mcg)	Lead found (mcg)	Difference (%)
1.00	V^{V}(50)	1.03	+3
1.00	Cr^{III}(50)	1.05	+5
1.00	Mo^{VI}(50)	0.96	−4
1.00	W^{VI}(50)	0.95	−5
1.00	Mn^{II}(100) Co^{II}(100) Ni(100)	1.00	0
1.00	Fe^{III}(100)	1.05	+5
1.00	Pd(100)	1.02	+2
1.00	Cu(100) Ag(100) Au^{III}(100)	1.03	+3
1.00	Zn(100) Cd(100) Hg(100)	1.00	0
1.00	Ga(50)	1.06	+6
1.00	In(100)	0.95	−5
1.00	Tl^{I}(100)	1.03	+3
1.00	Tl^{III}(10)	0.99	−1
1.00	Ge(100) Sn^{II}(100)	1.04	+4
1.00	As^{III}(100) Sb^{III}(100)	0.93	−7
1.00	Bi(10)	1.06	+6
1.00	Se^{IV}(100)	0.95	−5
1.00	Te^{IV}(10)	1.02	+2

RESULTS AND DISCUSSION

A number of samples with various known amounts of lead were analysed by the procedure described above. The amount of lead in standard lead diethyldithiocarbamate solution was 1.0 mcg/3 ml for the determination of 0.1–1.0 mcg of lead and 0.1 mcg/3 ml for the determination of 0.01–0.1 mcg of lead. The values of the ratio A_{aq}/A_{org} were plotted against x (cf.

Figure 1, curves A, B). The length of abscissa at the intersection is given by the blank experiment. The practical applicability of the method was tested by determining lead in the presence of a number of metals extractable as diethyl-dithiocarbamate. The results are summarized in Table I. From other elements only bismuth present in more than 10-fold excess interferes. Other elements do not interfere when present in 50–100-fold excess. The selectivity can be further increased by the preliminary separation of lead which need not be quantitative. The separation yield can be determined easily from the initial and final activity.

The proposed method is very simple, rapid, and selective. Because of the short half-life of ^{212}Pb (10.5 hr) the method is free from health hazards or laboratory contamination danger. Both the sensitivity and selectivity are much higher than those of the dithizone method.

References

1. G. Iwantscheff, *Das Dithizon und seine Anwendung in der Micro und Spurenanalyse* (Verlag Chemie, Weinheim, 1958).
2. J. Starý, *The Solvent Extraction of Metal Chelates* (Pergamon Press, Oxford, 1964).
3. S. M. Trascenko and E. V. Sobotovich, *Radiometric Methods for Determination of Microelements* (Nauka, Moscow, 1965).
4. J. Starý and R. Burcl, *Radiochem. Radioanal. Lett.* **7**, 235 (1971).
5. J. Starý, K. Kratzer, and A. Zeman, *J. Radioanal. Chem.* **5**, 71 (1970).
6. J. Růžička and J. Starý, *Substoichiometry in Radiochemical Analysis* (Pergamon Press, Oxford, 1968).

Determination of Lead, Cadmium, Copper and Zinc in Biological Materials by Anodic Stripping Polarography

I. ŠINKO and L. KOSTA

Boris Kidric Institute of Chemistry and Department of Chemistry, University of Ljubljana, Ljubljana, Yugoslavia

The high sensitivity of anodic stripping polarography and the simple equipment used make this technique very suitable for determining certain toxic elements, in particular lead and cadmium, as well as essential elements such as zinc or copper. These are found in biological and environmental systems in the concentration range of between nanograms per gram and a few hundred micrograms per gram of sample. The amount of sample required for one analysis is of the order of 100 mg, therefore the blank values introduced by the oxidizing mixture do not represent a serious limitation. After the decomposition of the sample by wet ignition no further separations are required. Copper and lead are determined from a solution made 1.0 M with respect to HCl. For zinc and cadmium the solution is buffered to pH 4.9–5.1.

Values are presented for a set of samples among which are standard kale, orchard leaves, and bovine liver. The uptake of lead, zinc, and cadmium has been measured in carrots grown in the environment of a lead-mining area near Mežica, Slovenia. The results are compared with those from a non-exposed site.

Anodic stripping polarography (ASP) has only recently been introduced as a technique for determining the two very toxic environmental contaminants lead and cadmium; some other elements, including copper and zinc, which belong to the essential element group, can also be determined using appropriate sample aliquots. The technique covers the concentration range of interest (10^{-8} M solutions) with adequate accuracy and precision and does not require expensive equipment.

Only a few papers dealing with the determination of trace metals in biological samples by ASP have been published so far. Both direct ashing[1-6] and wet ignition[7,8] have been used for the decomposition of the sample.

In the two last-mentioned methods, however, the elements are separated by extraction prior to their determination by ASP. In the present investigation it was shown that the technique can be applied directly to the solution obtained following wet ignition of the sample. By properly adjusting the medium, lead and copper can be determined from one aliquot of the solution and zinc and cadmium from another.

EXPERIMENTAL

Apparatus and Reagents

Current-potential curves were recorded with a polarograph (Type PO4, Radiometer, Copenhagen) at a rate of change of applied e.m.f. of 0.2 V/min. A sitting mercury drop (~ 6 mg) was used as indicator electrode. The volume of the electrolysis cell was 14 ml but the actual volume of the electrolysis solution was always 5.0 ml. The external reference electrode was a saturated calomel electrode (SCE) connected to the electrolysis solution by a salt bridge filled with agar-agar and 2 M potassium chloride solution. During electrolysis the solution was stirred with a glass stirrer at 400 rev/min. Oxygen was removed by bubbling pure argon ($O_2 < 0.01\%$) through the solution for 7 to 10 min.[9]

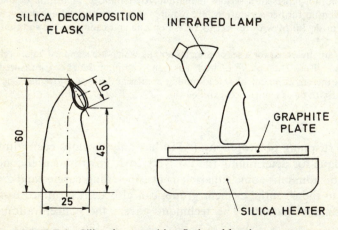

FIGURE 1   Silica decomposition flask and heating arrangement.

The amount of sample taken for one analysis was typically 100 mg. Decomposition was carried out in a silica container of about 10 ml as shown in Figure 1. Its upper part is narrow and bent in order to minimize losses by sputtering during the oxidation. The weighed sample was transferred into the container, the acid mixture was added, and then placed on a graphite plate fixed over a silica heater; the evaporation of the acid was assisted in the final stage by an additional infrared lamp (above), as shown in Figure 1.

All acids and sodium acetate used were of suprapure quality (Merck); H_2O_2 and NaOH were p.a.; water was first dionized and then doubly distilled from a silica still.

PROCEDURE

(a) Determination of Copper and Lead

The biological sample (~ 100 mg) is transferred into the silica decomposition flask, 0.5 ml nitric acid (d = 1.40 g/cm^3) and 0.2 ml sulphuric acid (d = 1.84 g/cm^3) are added, and the flask is heated on the graphite plate until all nitric acid has evaporated. The temperature is carefully increased until fumes of SO_3 start to form. Hydrogen peroxide (30%) is added down the wall of the flask until the dark solution clears. The total amount of peroxide required to complete the decomposition is between 0.5 and 0.8 ml.

The solution is finally evaporated to dryness. The residue is dissolved in 5 ml of warm 1 M HCl, taking care that all solids are removed from the walls by using a silica spatula. The cooled solution is transferred to the electrolysis cell and oxygen flushed out with argon.

The electrolysis is carried out at -0.8 V versus SCE for 3 to 10 min, depending on the concentrations of copper and lead in the solution. The current-potential curve is then recorded between -0.8 to 0.0 V versus SCE at a constant rate. The oxidation current peak for copper is at approx. -0.2 V and for lead at -0.4 V versus SCE, respectively. In sample with a high lead-to-copper ratio, resolution is improved if copper is electrolysed at -0.45 V (Figure 2).

The concentrations of copper and lead have been determined by the method of standard addition (the standard was added before the decomposition of the sample) as well as from calibration curves. A typical calibration curve is reproduced in Figure 3. The technique as described is applicable to concentrations of copper and lead as low as 10^{-7} M (0.006 mcg/ml) and 2.10^{-8} M (0.004 mcg/ml), respectively. At lower concentrations the accuracy and precision deteriorate due to reagent blanks. The amount of sample has to be increased if the concentrations of lead and copper in the material are lower than 0.2 ppm and 0.3 ppm, respectively.

Cadmium can also be determined from the same solution by carrying out the electrolysis at −0.85 V versus SCE; the cadmium maximum in this medium is approx. −0.6 V versus SCE. However, the accuracy and precision of results for cadmium are poorer due to the high reduction current of

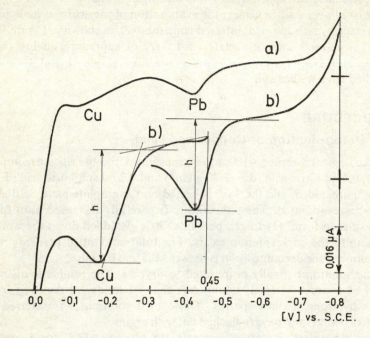

FIGURE 2 Current-potential curves for copper and lead ions in the acid solution. (a) blank; (b) solution containing 2.10^{-7} M Pb^{2+} and 2.10^{-6} M Cu^{2+}. Electrolysis time 3 min. Curve (a) electrolysis at −0.80 V versus SCE for lead and copper; curve (b) copper only electrolysis at −0.45 V versus SCE.

hydrogen ions (Figure 2). Weakly acid solutions are preferable. Therefore, the conditions for cadmium were modified to determine it together with zinc as described below.

(b) Determination of Cadmium and Zinc

A separate 100-mg aliquot of the sample is taken for the determination of zinc and cadmium. The same procedure as applied to copper and lead is followed, as far as the dissolution of the sulphuric acid residue in 5 ml of 1 M HCl is concerned. The solution is then transferred quantitatively into a 10-ml volumetric flask, 1–2 drops of methylorange are added, and the solution is carefully neutralized against this indicator, first with 2.5 N and later with 0.1 N NaOH and/or 0.1 N HCl. Finally the pH is adjusted to 5.3 by adding 0.5 ml of a 0.5 M sodium acetate-acetic acid buffer followed by addition of water to the mark. After mixing, 5 ml of the solution is pipetted into the electrolysis cell, oxygen is removed by bubbling argon through the solution for 7 to 10 min, and the solution is electrolysed for 3 to 10 min at −1.30 V

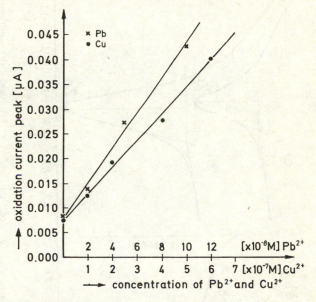

FIGURE 3   Calibration curves for copper and lead. Electrolysis time 10 min at −0.8 V versus SCE.

versus SCE for zinc and at −0.9 V versus SCE for cadmium, respectively. In most biological samples there is frequently a large excess of zinc with respect to cadmium. Therefore the current-potential curves are recorded separately for zinc and cadmium (Figure 4). The standard addition method as well as calibration curves have been used (Figure 5). Practical sensitivity limits for these two elements are 5×10^{-7} M (0.03 mcg/ml) for Zn and 10^{-8} M (0.001 mcg/ml) for Cd, equal to 3 ppm Zn and 0.1 ppm Cd when using 100 mg of sample for one determination. This limit can be further lowered for cadmium by prolonging the electrolysis time. In the case of zinc this is not feasible because of high blanks for this element. The amount of sample has to be increased adequately if the zinc concentration is lower.

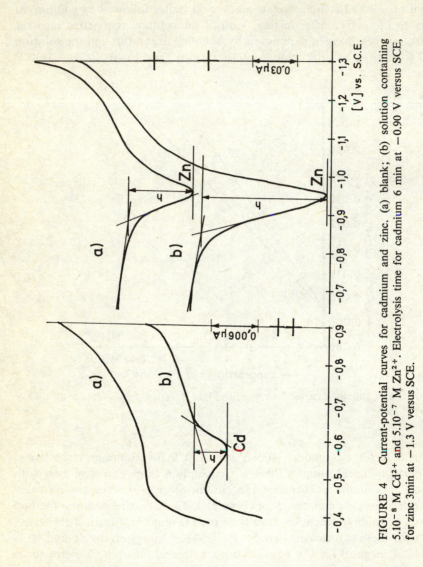

FIGURE 4 Current-potential curves for cadmium and zinc. (a) blank; (b) solution containing 5.10^{-8} M Cd^{2+} and 5.10^{-7} M Zn^{2+}. Electrolysis time for cadmium 6 min at -0.90 V versus SCE, for zinc 3 min at -1.3 V versus SCE.

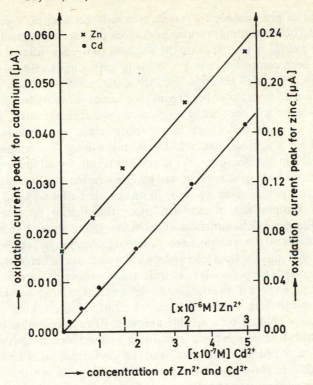

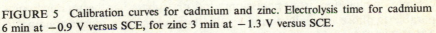

FIGURE 5   Calibration curves for cadmium and zinc. Electrolysis time for cadmium 6 min at −0.9 V versus SCE, for zinc 3 min at −1.3 V versus SCE.

RESULTS AND DISCUSSION

For accurate determinations of metals in biological materials quantitative mineralization of the sample is very important. Data in the literature indicate that results are low for all four elements if ashing is used for mineralization.[10] According to Doshi *et al.*[11] losses for zinc and cobalt occur already at the ignition temperature of 400°C. Therefore, wet ashing is preferable provided that the reagents used are extremely pure and that blanks are determined or known.

High cathodic currents were observed if the decomposition was not complete. Best results were obtained when using a mixture of sulphuric and nitric acids and assisting the oxidation by dropwise addition of H_2O_2 into hot solution. Final solutions were either clear or there was a small amount of solids (calcium sulphate or silica) left when following this procedure. Perchloric acid is not suitable for plant materials because of a considerable precipitation of the sparingly soluble potassium perchlorate formed.

It is also necessary to evaporate the solution to dryness after the mineralization has been completed in order to avoid cathodic currents at somewhat lower potentials, which interfere with the determination of zinc. These are very likely due to traces of either organic substances or of hydrogen peroxide.

The choice of the supporting electrolyte is another critical parameter. Newberg and Christian[2] claim that copper, lead, and cadmium can be determined in biological materials by simply ashing the sample and dissolving the ash in distilled water. We found low results for all four components if either ashing or wet ignition, followed by evaporation, was used and then the residue taken up in distilled water or in an acetate buffer solution pH 5–6.5. The latter medium was investigated more thoroughly because it would allow simultaneous determination of all four components.[12] In a series of experiments carried out with pure solution to which sulphate ions were added, we found low results for both lead and zinc in a medium buffered with acetate. Replacing acetate by potassium chloride leads to loss of accuracy for lead and copper. Therefore it was decided to determine the latter two elements in 1 M hydrochloric acid which dissolves practically all the residue. Biological samples always contain considerable amounts of calcium, magnesium, and iron; neutralization of solutions containing hydrochloric acid (prior to the determination of cadmium and zinc) leads to the formation of hydroxyl salts or hydroxides. Experiments with synthetic solutions to which amounts of calcium, magnesium, iron, potassium, and phosphate had been added comparable to those normally found in this type of sample (corresponding to 3% Ca, 1% Mg and 0.1% Fe, 2% K and 1.6% P in the original sample) have shown that a precipitate does not form if the pH does not exceed 5.5. The lower pH limit, on the other hand, is given by the high reduction current of hydrogen ion which appears in solutions of pH less than 4.7; this interferes with the measurement of the oxidation current peak for zinc (Figure 4). This interference begins to appear at pH 5.3 if the concentration of zinc is less than 5×10^{-7} M, but this was not the case with any of the samples analysed so far.

Using the above procedure, solutions could be analysed 10^{-8} M with respect to lead, copper, and zinc provided that the blank values for these elements were negligible. In the case of cadmium there is a linear dependence of maximum anodic current with concentration in the broad range from 10^{-8} M to 10^{-5} M. Of other ions mentioned (Ca, Mg, Fe, K, HPO_4^-) only the last-mentioned affects the anodic maxima of cadmium (Figure 6) as well as of zinc in solution at pH 5, but does not influence lead and copper peaks in 1 M hydrochloric acid. In order to minimize the influence of the varying concentration of these interferences present in different biological samples on the

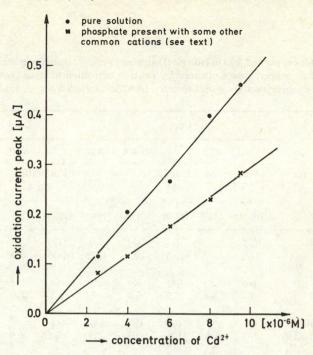

FIGURE 6   Influence of phosphate on the calibration curves for cadmium ion in the concentration range $2.5.10^{-6} - 9.5.10^{-6}$ M. Electrolysis time 3 min at -0.9 V versus SCE.

accuracy of the results, they have all been analysed by the use of calibration curves as well as by the method of standard addition. The data presented in Tables I and II for eight different biological samples indicate good agreement between the two approaches. The results all appear to be within the limits of precision achievable by using ASP.

For some of the materials in the tables, results obtained by other techniques are included for comparison and show good agreement. The first five samples were in the form of homogeneous powder. The last three were obtained from samples cut to cubes of approx. 5 mm edge, and freeze-dried. The lower precision reflects the inhomogeneity of these samples due to differences in concentration within different sections of the plant. (Compare the rel. S.D. in Tables I and II for the freeze-dried carrot sample with the other three carrot samples.)

The high concentration of lead and arsenic in orchard leaves probably indicates that they might have been sprayed at some stage with preparations containing these two elements. It is known that arsenic(III) interferes in lead

TABLE I

Results for copper and lead in biological samples by anodic stripping polarography (ASP). For copper, values obtained by neutron activation analysis (NAA) and atomic absorption spectrophotometry (AA) are included for comparison

Sample	Copper					Lead		
	ASP			NAA	AA	ASP		
	Calibration curve	(%)[a]	Stand. add.			Calibration curve	(%)[a]	Stand. add.
	(ppm)	(%)[a]	(ppm)	(ppm)	(ppm)	(ppm)	(%)[a]	(ppm)
Orchard leaves[b]	(7)[c] 11.2	9.2	(3) 12.7	(5) 10.8	(7) 12.9	(10) 44.0	5.2	(3) 43.4
Tomato leaves[b]	(6) 7.7	6.6	(4) 7.7	(5) 9.4		(6) 4.5	2.8	(3) 3.2
Bovine liver,[b] freeze-dried	(6) 197	6.4	(6) 195	(9) 191	(8) 205	(4) 0.36	6.9	(4) 0.39
Kale[d]	(6) 5.2	11.8	(4) 5.0		(4) 4.3	(7) 2.3	10.3	(3) 2.9
Carrot, freeze-dried	(8) 8.4	4.9	(3) 8.6			(6) 1.2	10.5	(4) 1.1
White carrot	(5) 2.4	12.7	(2) 2.3			(4) 0.57	23.3	(3) 0.71
Carrot 3	(7) 4.0	15.3	(6) 4.1			(6) 15.3	19.3	(5) 15.4
Carrot 6	(7) 3.0	11.6	(3) 3.3			(6) 3.4	23.5	(3) 3.6

a Rel. S.D.

b These samples were kindly supplied by the U.S. National Bureau of Standards.

c Figures in parentheses give the number of determinations.

d This sample is an international reference sample kindly supplied by Professor H. J. M. Bowen.

TABLE II

Compositive results for zinc and cadmium in biological samples by ASP. For zinc, values obtained by AA are included

Sample	Zinc				Cadmium		
	ASP			AA	ASP		
	Calibration curve	$(\%)^a$	Stand. add.		Calibration curve	$(\%)^a$	Stand. add.
	(ppm)		(ppm)	(ppm)	(ppm)		(ppm)
Orchard leaves[b]	$(7)^c$ 28.3	9.3	(3) 29.8	(5) 29.5	(6) < 0.1		(3) < 0.1
Tomato leaves[b]	(6) 62.9	2.7	(6) 62.9	(5) 61.0	(6) 2.4	9.1	(5) 2.1
Bovine liver,[b] freeze-dried	(7) 118	3.7	(4) 119	(5) 124	(5) < 0.1		(5) < 0.1
Kale[d]	(6) 29.4	5.2	(3) 28.5		(6) 0.75	10.3	(6) 0.72
Carrot, freeze-dried	(7) 21.3	4.9	(3) 20.8		(5) 0.68	10.5	(2) 0.74
White carrot	(4) 12.4	26.0	(2) 12.1		(4) 0.31	13.3	(4) 0.28
Carrot 3	(6) 41.6	4.5	(3) 39.9		(6) 1.7	17.6	(5) 1.5
Carrot 6	(6) 31.5	4.5	(2) 30.2		(6) 1.3	24.5	(4) 1.6

[a] Rel. S.D.

[b] These samples were kindly supplied by the U.S. National Bureau of Standards.

[c] Figures in parentheses give the number of determinations.

[d] This sample is an international reference sample kindly supplied by Prof. N. J. M. Bowen.

determinations due to the same half peak potential during its oxidation. However, following wet ignition of the sample arsenic was expected to be in the pentavalent state. To prove this the effects of arsenic(III) added to the oxidizing mixture were investigated.

Measurements of the oxidation current peak of 8×10^{-6} M lead solution alone and of the same solution containing as much as 10^{-4} M added As(III) before treatment with the oxidants, and then following the whole procedure, were 0.344 μA and 0.355 μA, respectively, thus showing evidence of negligible interference. At the potential used during electrolysis there is evidently no reduction of As(V). Compared to lead, pentavalent arsenic has its anodic maximum at a much more positive potential, but since it is not reduced under the conditions used it does not affect the accuracy of lead determination.

Using this method, we could clearly demonstrate not only considerable differences in heavy metal concentration in plant tissues but also a strong dependence upon the area where they were grown. Although concentrations of copper in carrot samples do not vary essentially, lead, zinc and cadmium differ in a much wider range and are particularly high in samples 3 and 6 which were grown close to a lead zinc mine. The precision and accuracy make the technique suitable for environmental studies involving uptake studies and distribution measurements.

Acknowledgements

This paper summarizes part of the work supported by Grant No. NBS/G/-107, U.S. Department of Commerce, National Bureau of Standards, which is gratefully acknowledged.

References

1. M. Ariel and U. Eisner, *Isr. J. Chem.* **1**, 295 (1963).
2. C. L. Newberg and G. D. Christian, *J. Electroanal. Chem.* **9**, 468 (1965).
3. I. V. Markova and S. J. Sinyakova, *Agrokhimiya* **12**, 118 (1966). [*Anal. Abstr.* **15**, 2321 (1968).]
4. R. Neeb, *Inverse Polarographie und Voltammetrie* (Verlag Chemie, Weinheim, 1969).
5. V. D. Melekhin and E. M. Roizenblat, *Lab. Delo* **2**, 107 (1969). [*Anal. Abstr.* **18**, 3251 (1970).]
6. W. Oelschlaeger and R. Gilg, *Landwirt. Forsch. Sonderh.* **22**, 218 (1969). [*Anal. Abstr.* **20**, 1821 (1971).]
7. H. K. Hundley and E. C. Warren, *J. Ass. Offic. Anal. Chem.* **53**, 705 (1970).
8. B. Morches and G. Tölg, *Z. Anal. Chem.* **250**, 81 (1970).
9. I. Šinko and J. Doležal, *J. Electroanal. Chem.* **25**, 53 (1970).
10. T. T. Gorsuch, *The Destruction of Organic Matter* (Pergamon Press, Oxford, 1970).
11. G. R. Doshi, C. Sreekumaran, C. D. Mulay and B. Patel, *Curr. Sci.* **38**, 206 (1969).
12. I. Šinko and J. Doležal, *J. Electroanal. Chem.* **25**, 299 (1970).

Determination of Trace Amounts of Antimony by Flameless Atomic Absorption Spectroscopy

B. E. SCHREIBER and R. W. FREI†

Analytical Department of the Pharmaceutical Division,
Sandoz Ltd., 4002 Basle, Switzerland

KEY WORDS: antimony, flameless atomic absorption spectroscopy, titaniun dioxide

A method for the determination of traces of antimony in organic and inorganic material is described. One gram of the sample is decomposed with sulphuric acid and hydrogen peroxide in a closed system; hydrochloric acid is added and antimony is extracted from the solution with diisopropyl ether. Aliquots of the extract are atomized in a heated graphite tube for atomic absorption measurement. Three possibilities of antimony measurement with detection limits of 0.6, 0.06, and 0.02 mcg per 1 g sample (1 % absorption) are described. The method is suitable for routine work and compares favourably to flame and spectrophotometric methods.

INTRODUCTION

The availability of good analytical methods for the trace analysis of antimony is of prime importance, due to its high toxicity even at trace concentration levels (ppb). The determination of low levels of antimony in food dyes, printing inks and laboratory chemicals is a problem that has not yet been solved satisfactorily. The classical methods, by extraction, complex formation and spectrophotometric evaluation,[1-6] are subject to many interferences. In addition, these methods are not applicable to concentrations below 1 mcg.

† On leave from Dalhousie University, Halifax, Nova Scotia, Canada.

Flame atomic absorption spectroscopy gives improved selectivity, but the sensitivity is not sufficient to estimate antimony at the 1-ppm level from 1 g sample.[7,8] Solvent extraction, followed by direct aspiration of the extract into the flame,[9-12] does not increase sensitivity sufficiently. Detection limits can be improved by the use of an air hydrogen flame.[13] Atomic fluorescence spectroscopy yields better results for a direct measurement of the sample solutions; 0.5 mcg of antimony was given as the minimum detectable amount.[14,15] With the introduction of flameless methods impressive improvements in sensitivity have been reported for antimony,[16,17] with detection limits as low as 5.10^{-12} g. Flameless techniques have been adopted with good success in our laboratories for the analysis of printing inks and TiO_2-containing samples. Some of the problems and results encountered in developing a technique for antimony are described in this paper.

EXPERIMENTAL

Instrumentation

Decomposition of the material was carried out in a closed reflux system described by Gorsuch.[18]

A Perkin-Elmer Model 403 atomic absorption spectrophotometer equipped with a deuterium background compensator and a graphite furnace HGA 70 together with a Perkin–Elmer hollow cathode lamp and a PE 165 recorder were used. The graphite oven has a magnetic valve to stop the inert gas flow during the measurement. Samples (50 mcl) were injected into the graphite oven by Oxford pipettes (Oxford Laboratories, California) with disposable polypropylene tips.

The samples were dried for 60 sec at 100°C, heated for 60 sec 800°C and atomized for 20 sec at 2200°C. Argon was used as inert gas to protect the graphite tube from oxidation.

Instrument settings   The spectral line at 217.58 nm was used with a slit width of 0.3 mm (2 Å resolution). Twenty-five per cent absorption was recorded full-scale. Deuterium background compensation was used for all measurements.

Spectrophotometric measurements of antimony were carried out with a Pye-Unicam spectrophotometer SP 600.

Reagents

All chemicals were of analytical-grade purity. All solutions were made from doubly distilled water. A stock solution, containing 1000 ppm of antimony, was prepared by dissolving 2.743 g of antimony tartrate hemihydrate

(Merck, Darmstadt, DBR) K SbO $C_4H_4O_6 \cdot 1/2 \, H_2O$ in 200 ml 50% (v/v) sulphuric acid, and diluting to one litre in a volumetric flask. Dilution series were prepared with 1% (v/v) sulphuric acid.

Procedure

Decomposition of the material The sample (1 g, organic or TiO_2) was heated in the decomposition apparatus[18] with 10 ml of 50% (v/v) sulphuric

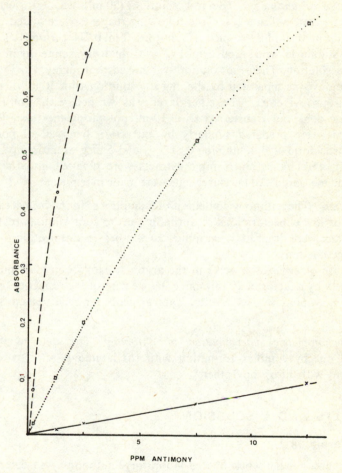

FIGURE 1   Calibration curves for antimony after extraction.

————: Single injection of 50 mcl of the extract into the cuvette.

– – – –: Four aliquots of 50 mcl of the extract injected into the cuvette, argon flow stopped during atomization.

..............: Single injection of 50 mcl of the extract, argon flow stopped during atomization.

acid. The water was distilled into the reservoir and transferred to a 50-ml measuring flask. The sulphuric acid was then refluxed for a few minutes. After cooling, 0.5 ml of hydrogen peroxide was added and the solution was heated until sulphuric acid fumes appeared. This step was repeated until all organic and inorganic matter had dissolved. The solution was transferred to a 50-ml flask with 32 ml of concentrated hydrochloric acid and diluted to volume with water.

Extraction of antimony   The test solution (20 ml) was transferred to a 100-ml separatory funnel, and the following reagents were added: 1 ml 1% (wt) sodium disulphite, aqueous solution; 2 ml 0.2N cerium(IV) sulphate in 1N sulphuric acid; and 2 ml 1% (wt) hydroxylamine hydrochloride, aqueous solution. Then the solution was mixed thoroughly.

Antimony was separated by shaking the solution with 10 ml of peroxide-free diisopropyl ether. The ether layer was washed with 2 ml 10% (wt) sulphuric acid and transferred to a 10-ml polypropylene tube. Reference standards were prepared similarly by extraction from 20-ml portions of 6N hydrochloric acid containing 0, 0.5, 1, and 5 mcg of antimony. Aliquots (50 mcl) of reference and sample extracts were pipetted into the graphite oven and measured. All measurements were made in triplicate.

Evaluation   The antimony content of the sample was found from a calibration plot. For low concentrations of antimony, up to four aliquots of 50 mcl of the extract were dried in the graphite cuvette before atomization to improve the detection limit.

The use of a gas stop valve in the argon stream[19] can also improve the sensitivity by a factor of 10 through enhancement of the residence time of the antimony vapour in the cuvette. Typical calibration curves are shown in Figure 1.

Spectrophotometric determination of antimony   The determination was carried out by complex formation with rhodamine B after extraction of antimony with diisopropyl ether.[6]

RESULTS AND DISCUSSION

Interferences

During preliminary work in this laboratory, antimony was determined in aqueous solutions which contained high concentrations of hydrochloric, nitric, and sulphuric acid and alkali, iron, and titanium salts; this resulted in serious interferences. With the present procedure, however, interferences from high concentrations of hydrochloric, nitric, and sulphuric acids and salts in the test solution were eliminated. Figure 2 shows the effect of acid concentration on the antimony signal. Water and very dilute acids (less than 0.1%) gave low results due to hydrolysis. Therefore the lowest acid concentration employed was 0.1%.

The presence of hydrochloric acid resulted in low measurements, possibly due to formation of volatile antimony halides; nitric acid caused high values at low concentrations, but showed a strong variation of the values with varying acid amounts. Sulphuric acid gave optimal results. Only small differences in antimony recovery were observed with concentrations of sulphuric acid between 0.1 and 10% (wt). Of the interfering cations, iron and titanium were tested in high concentrations. In Figure 3 it can be seen that in hydrochloric acid solution, interferences are very strong. Addition of 1% (v/v) sulphuric acid to the test solution removes the interference only partially. A direct estimation of antimony in titanium compounds was therefore not possible.

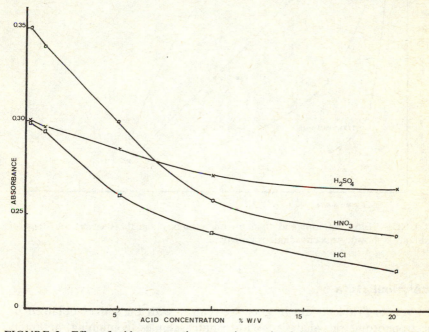

FIGURE 2   Effect of acid concentration on antimony signal. 20 ng of antimony measured. Aqueous solutions were used.

Methylisobutyl ketone, benzene, and diisopropyl ether have been tested for extraction of antimony from hydrochloric acid solutions. Diisopropyl ether gave the best results. Extraction was carried out from 6N hydrochloric acid medium and antimony was determined by atomizing 50-mcl aliquots of the extract. In this procedure, the major problem was the volatility of antimony in the presence of hydrochloric acid during the evaporation step in the graphite tube. This could be eliminated by washing the extract with a small amount of sulphuric acid prior to atomization. Back extraction of antimony by sulphuric acid washing is very small and does not influence antimony values.

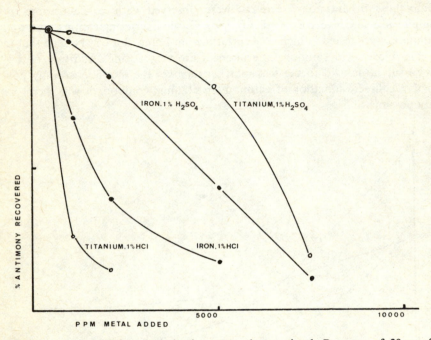

FIGURE 3   Effect of iron and titanium on antimony signal. Recovery of 20 ng of antimony is shown. Aqueous solutions were used.

Analytical Data

The results (Table I) are in good agreement with the values obtained by spectrophotometric analysis. With small samples containing antimony concentrations below 10 ppm, measurement by the proposed method will give more reproducible results than flame methods or spectrophotometric measurements. For 1 g sample, the detection limit was found at 0.6 ppm (= 12 ng of antimony atomized for a 1% absorption signal) using the standard procedure. With the argon flow stop method, the sensitivity can be improved and a detection limit of 0.06 ppm (1.2 ng antimony atomized) has been observed.

TABLE I

Results of antimony determination with a graphite furnace compared to a spectro-photometric assay[a]

Sample material	Sample size	ppm Sb HGA 70	ppm Sb spectro-photometric
TiO_2 I	1 g	9.5 $\pm$0.5	9.4$\pm$1.3
TiO_2 II	1 g	12.2 $\pm$0.6	10.6$\pm$1.2
TiO_2 III	0.2 g	42$\pm$2	45$\pm$3
$TiCl_3$, 15% solution	5 ml	1.6 $\pm$0.1	2.3$\pm$0.5
Printing ink (20% TiO_2)	1 g	3.5 $\pm$0.2	3.8$\pm$0.6
Hydrochloric acid (reagent grade)	20 ml	0.03$\pm$0.01	not detected

[a] Mean values and S.D. of eight determinations are listed. Single injection procedure was used.

Injection of four aliquots of 50 mcl into the graphite tube in combination with argon flow stop yielded a detection limit of 0.02 ppm. With these three possibilities of measurement, the method can easily be fitted to a wide range of antimony concentrations without altering the procedure.

Due to the preparation of the sample in a closed system, no antimony losses occurred even with materials with high halogen concentration or high amounts of organic material. Interferences by other elements were minimized by the extraction step and the high selectivity of atomic absorption measurement. The method can be considered suitable for routine work.

Acknowledgements

The helpful discussions and assistance of Mr. H. D. Seltner, Sandoz Ltd., Basle, and of Drs. T. F. Bidleman and R. Stephens, Dalhousie University, Halifax, were much appreciated.

References

1. V. Stresko and E. Martiny, *At. Absorption Newslett*. **11**, 4 (1972).
2. R. A. Mostyn and A. F. Cunningham, *Anal. Chem.* **39**, 433 (1967).
3. A. G. Fogg, C. Burgess, and D. Thorburn Burns, *Talanta* **18**, 1175 (1971).
4. T. H. Maren, *Anal. Chem*. **19**, 487 (1947).
5. V. Stara and J. Stary, *Talanta* **17**, 341 (1970).
6. F. M. Ward and H. W. Lakin, *Anal. Chem*. **26**, 1168 (1954).
7. A. A. Yadav and S. M. Khopkar, *Bull. Chem. Soc. Jap*. **44**, 693 (1971).
8. D. J. Nicolas, *Anal. Chim. Acta* **55**, 59 (1971).
9. J. C. Meranger and E. Somers, *Analyst* **93**, 799 (1968).
10. M. Yanagisawa, M. Zuzuki, and T. Takeuchi, *Anal. Chim. Acta* **47**, 121 (1969).
11. K. E. Burke, *Analyst* **97**, 19 (1972).

12. J. B. Headridge and D. Risson Smith, *Lab. Pract.* **20**, 312 (1971).
13. K. Fuwa and L. Valee, *Anal. Chem* **35**, 942 (1963).
14. R. M. Dagnall, K. C. Thompson, and T. S. West, *Talanta* **14**, 1151 (1967).
15. A. Fulton, K. C. Thompson, and T. S. West, *Anal. Chim. Acta* **51**, 373 (1970).
16. B. V. L'Voy, *Zh. Anal. Khim.* **26**, 510 (1971).
17. D. A. Katskov and B. V. L'Vov, *Zh. Prikl. Spectrosk.* **10**, 382 (1969).
18. T. T. Gorsuch, *Analyst* **84**, 135 (1959).
19. H. L. Kahn and S. Slavin, *At. Absorption Newslett.* **10**, 125 (1971).

Determination of Some Transition Metals by Atomic Absorption Spectroscopy after Extraction with Pyridine-2-aldehyde-2-quinolylhydrazone†

R. W. FREI, T. BIDLEMAN, G. H. JAMRO, and O. NAVRATIL

Department of Chemistry, Dalhousie University
Halifax, Nova Scotia, Canada

The extraction of pyridine-2-aldehyde-2-quinolylhydrazone chelates of cadmium, cobalt, copper, nickel, and zinc into isoamyl alcohol (IAOL) and methylisobutyl ketone (MIBK) has been investigated as a basis for the determination of these metals. Below pH 6 the extraction is enhanced by the addition of perchlorate, suggesting that charged complexes are being extracted by ion-pair formation. Sensitivities (1% absorption) are reported for IAOL and MIBK solutions of the metals sprayed into an air–acetylene flame. A procedure for the determination of the above metals by atomic absorption spectroscopy after extraction is given. The procedure is applied to the analysis of tap water for cadmium, copper, and zinc.

INTRODUCTION

Substituted heterocyclic hydrazones form chelates with a number of transition metals which are extractable into immiscible organic solvents. Because of the high molar absorptivities of the resulting chelates, spectrophotometric techniques have been applied until now. Pyridine-2-aldehyde-2-quinolylhydrazone (PAQH) has been used for the photometric determination of palladium,[1,2] cobalt, and nickel.[3,4] Quinoline-2-aldehyde-2-quinolylhydra-

† Presented at the National Conference of the Canadian Institute of Chemistry, Halifax, N.S., May/June 1971.

zone (QAQH) has found use for the determination of copper,[5,6] and zinc has been determined with phenanthridine-6-aldehyde-2-quinolylhydrazone.[7] Under controlled conditions the procedures developed were specific for the metals involved; the use of masking agents and/or careful pH control was necessary because of the similarity of the chelate spectra.

In this work attention is focused on the atomic absorption determination of cadmium, cobalt, copper, nickel, and zinc after extraction with PAQH. An investigation of the extractability of the PAQH chelates of the above metals has been made by three of the authors.[8]

EXPERIMENTAL

Reagents and apparatuses

Solutions of cobalt, copper, nickel, and zinc were prepared from Fisher Certified Atomic Absorption Standards; cadmium solutions were prepared by dissolving the reagent-grade metal in dilute nitric acid. Reagent-grade or redistilled solvents were used. PAQH was prepared as previously described;[9] a 1% solution in 0.05 M hydrochloric acid was stable indefinitely. Solutions of the reagent in organic solvents were freshly prepared daily. All other chemicals were of reagent-grade quality.

Spectra were measured with Bausch & Lomb Spectronic 505 and Unicam 8000 Spectrophotometers. A Beckman 1301 DB-G Atomic Absorption Spectrophotometer was used for atomic absorption measurements. Perkin–Elmer single-element hollow cathode lamps were used for cobalt, copper, cadmium, and nickel; a calcium–zinc multielement lamp was used for zinc. Measurements of pH were made with a Radiometer 28 pH meter.

Procedure

a) *Spectrophotometric work*   Aqueous solutions of the metals containing PAQH, acetate buffer, and sodium perchlorate were shaken with an approximately equal volume of organic solvent alone or containing PAQH. Equilibrations were done in separatory funnels or test tubes with ground glass stoppers, using a mechanical shaker or home-made rotator. Manual shaking was used for short equilibration times. Equilibrium was established within two minutes for those systems in which the reagent was added as an aqueous solution in dilute hydrochloric acid, while longer shaking times (3–4 hours) were required when the reagent was added as a solution in the organic solvent. An aliquot of the extract was taken for spectrophotometric measurements.

b) *Interference study and tap water analysis*   Aqueous solutions of the metals (100 ml) containing 0.08 mmole of PAQH (added as a 1% solution in

0.05 *M* HCl), 0.5 mmole of sodium perchlorate, and acetate buffer were shaken with 10 ml of isoamyl alcohol (IAOL) or methylisobutyl ketone (MIBK) for two minutes. A pH of 5–6 was used for the extraction of copper, cobalt, and nickel; zinc and cadmium were extracted at pH 6–8. After allowing the phases to separate for 30 minutes, the extracted metals were determined by atomic absorption. The MIBK extracts were directly aspirated; the IAOL extracts were diluted with an equal volume of ethanol before aspirating. Calibration curves were prepared by extracting known quantities of the metals under the same conditions.

RESULTS AND DISCUSSION

The optimum pH ranges for the extraction of the metals under the conditions used for the water analyses are shown in Figures 1 and 2. It was found that if perchlorate is not present, the recovery of all metals is considerably less at pH < 6 (MIBK solvent). This suggests that in the lower pH region the metals are extracted partially or wholly as ion-pair complexes with perchlorate, and

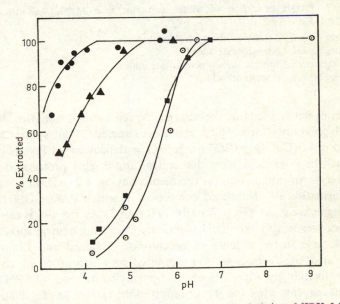

FIGURE 1   PAQH extraction of copper, cadmium, and cobalt into MIBK 0.08 mmole PAQH, V_w = 100 ml, V_0 = 8 ml.

● 15 mcg copper, 0.5 mole sodium perchlorate added
○ 15 mcg copper, no perchlorate added
■ 6 mcg cadmium, 0.5 mmole sodium perchlorate added
▲ 45 mcg cobalt, 0.5 mmole sodium perchlorate added

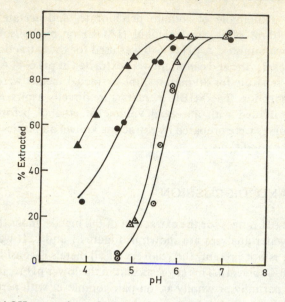

FIGURE 2 PAQH extraction of nickel and zinc into MIBK 0.08 mmole PAQH, $V_w = 100$ ml, $V_o = 8$ ml.

▲ 40 mcg nickel, 0.5 mmole sodium perchlorate added
△ 40 mcg nickel, no perchlorate added
● 5 mcg zinc, 0.5 mmole sodium perchlorate added
○ 5 mcg zinc, no perchlorate added

the chelates must therefore be charged species. Such behavior has been observed previously for cobalt,[8] and the extracted cobalt species has been proposed to be $(Co^{III}Q^+)(ClO^-)$, where Q = deprotonated PAQH (HQ and H_2Q^+ will be used to indicate the neutral and doubly protonated reagent, respectively). An explanation for the metals having $+2$ oxidation states may be the formation of protonated complexes in which PAQH is acting as a neutral ligand, e.g. $M^{II}(HQ)_2^{2+}$ or $M^{II}(HQ)(Q)^+$. As the pH is raised these complexes presumably become deprotonated, forming neutral species which are extractable in the absence of perchlorate. Geldard and Lions[10] have demonstrated the formation of protonated and deprotonated complexes for the ligand pyridine-2-aldehyde-2-pyridylhydrazone (PAPHY) and analogous reagents. Heit and Ryan[1] reported the extraction of palladium into chloroform as $Pd(HQ)SO_4$; however, in an earlier paper[9] on PAQH chelates they reported no evidence for protonated complexes.

The visible spectra of the extracted metal complexes are similar, having maxima between 480 and 520 nm. The reagent absorbs only slightly in this region. Spectro-photometric sensitivities are considerably less for cadmium and zinc than for the other metals under the conditions used in the water analyses, thus atomic absorption offers a real advantage for the determination of these two metals after PAQH extraction. The effective molar absorptivities of the chelates in MIBK depends upon the pH of the solution from which they were extracted, and for this reason they are not reported here. Figure 3 shows an example of this behaviour for the nickel–PAQH system. Under the experimental conditions the nickel was 95–100% extracted (even at pH 4.6) because of the larger solvent:water ratio than was used in the analytical procedure. This pH dependence further suggests the formation of different species in the low and high pH ranges.

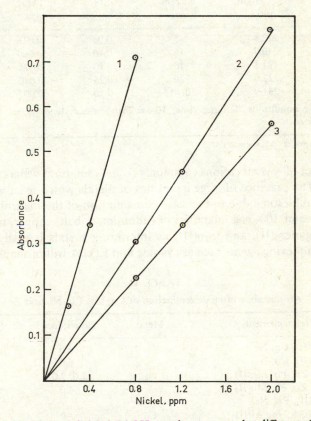

FIGURE 3   Absorbance of nickel–PAQH complexes extracted at different pH values.
$V_w = 8$ ml, $V_0 = 5$ ml, 0.0064 mmole PAQH
1. no perchlorate, pH 6.7
2. 0.04 mmole sodium perchlorate added, pH 5.0
3. 0.04 mmole sodium perchlorate added, pH 4.6

Of the solvents used in this work, MIBK is the most suitable for atomic absorption work. The more viscous solvent IAOL should be diluted with ethanol before aspiration and has the added disadvantage of having a higher solubility in water than MIBK. Although the chelates are extractable into benzene, this solvent is unsuitable for atomic absorption work because it produces a smoky yellow flame. Atomic absorption sensitivities (1 % absorption) are given in Table I.

TABLE I

Sensitivities (1 % absorption) for atomic absorption determination

Metal	Wavelength nm	Lamp current ma	Slit mm	IAOL ppm[a]	MIBK ppm
Cd	228.8	5	0.25	0.016	0.012
Co	240.7	15	0.10	0.060	0.045
Cu	324.8	10	0.10	0.025	0.040
Ni	232.0	20	0.25	0.040	0.083
Zn	231.9	10—15	0.50	0.007	0.007

Working conditions: laminar flow, 10-cm lean air-acetylene flame, cold mode, single pass

[a]In the presence of 50% v/v ethanol.

The effect of several cations commonly found in natural waters is shown in Table II. The presence of large quantities of metals which react with PAQH may be troublesome due to excessive consumption of the reagent. However, the presence of 100-mcg quantities of cadmium, cobalt, copper, nickel, lead, zinc, manganese(II), and iron(III) as interfering metals is easily tolerated. Strong complexing agents such as cyanide and EDTA will probably interfere.

TABLE II

Atomic absorption determination of Cd, Co, Cu, Ni, and Zn

Foreign elements	Metal	Taken, mcg	Found, mcg
100 mcg Fe(III), Co, Ni		4.0	4.2
200 mcg Fe(III), Co, Ni		6.0	6.6
100 mcg Cu, Zn, Pb, Mn(II)		2.0, 2.0	2.2, 2.0
200 mcg Cu, Zn, Pb, Mn(II)		4.0	4.5
100 mcg Fe(III), Co, Ni, 1.3 mg Cr(III), 1.7 mg Al(III)	Cd	2.0	1.8
100 mcg Fe(III), Co, Ni, Cu, Zn, Pb, Mb(II)		4.0	4.0
1.1 mg Mg, 1.3 mg Cr(III), 1.3 mg Ca, 1.7 mg Al(III)		6.0	6.1

TABLE II—cont.

Foreign elements	Metal	Taken, mcg	Found, mcg
none		3.0	3.4
		6.0	5.4
100 mcg Fe(III), Cu, Ni, Pb, Mn(II)		0.0	0.0
		4.0	4.1
100 mcg Fe(III), Cu, Cd, Ni, Pb, Zn, Mn(II)		0.0	0.0
		2.0	2.0
	Co	4.0	4.4
200 mcg Fe(III), Cu, Cd, Ni, Pb, Zn, Mn(II)		2.0	2.0
		4.0	4.2
		6.0	6.7
100 mcg Fe(III), Cu, Cd, Ni, Pb, Zn, Mn(II), 0.9 mg Al(III), 1.3 mg Cr(III), 1.5 mg Ca, 1.1 mg Mg		2.0	1.8
		4.0	4.2
		10.0	11.0
100 mcg Fe(III), Co, Cd, Ni, Pb, Zn, Mn(II)		4.0	4.2
200 mcg Fe(III), Co, Cd, Ni, Pb, Zn, Mn(II)		12.0	11.6
1.1 mg Mg, 1.3 mg Ca 1.3 mg Cr(III), 1.7 mg Al(III)		8.0	8.3
		8.0	8.3
1.1 mg Mg, 1.3 mg Cr(III), 1.7 mg Al(III)		4.0	4.0
100 mcg Fe(III), Co, Ni, Cd	Cu	8.0	8.0
200 mcg Pb, Zn, Mn(II)		12.0	11.7
100 mcg Cd, Ni, Pb, Zn, Mn(II)		4.0	4.2
100 mcg Fe(III), Co, Ni, Pb, Zn, Mn(II), 1.1 mg Mg, 1.3 mg Cr(III), 1.3 mg Ca, 1.7 mg Al(III)		0.0	0.0
		12.0	12.2
1.1 mg Mg, 1.3 mg Ca 1.3 mg Cr(III), 1.7 mg Al(III)		10.0	10.5
		20.0	18.5
200 mcg Fe(III), Co, Cd, Cu, Pb, Zn, Mn(II)	Ni	10.0	10.0
300 mcg Cd, Pb, Zn, Mn(II)		20.0	19.5
500 mcg Fe(III), Co		10.0	10.0
200 mcg Fe(III), Co, Cd, Cu, Pb, Zn, Mn(II), 1.1 mg Mg, 1.3 mg Ca, 1.3 mg Cr(III), 1.7 mg Al(III)		0.0	0.0
		15.0	13.5
1.7 mg Al(III)		2.0, 2.0	1.9, 1.9
1.1 mg Mg, 1.3 mg Ca, 1.3 mg Cr(III), 1.7 mg Al(III)		1.0	1.0
100 mcg Cd	Zn	2.0	2.0
200 mcg Pb		3.0	3.0
200 mcg Mn(II)		3.0	3.0
200 mcg Fe(III), Co, Ni		3.0	2.7

Application of the method

Filtered laboratory tap water was analyzed for cadmium, copper, and zinc by the above procedure, using MIBK as the extracting solvent. Results are given in Table III. The cadmium content of the tap water was below the detection limits of this procedure.

TABLE III

Analysis of tap water[a] by atomic absorption after PAQH extraction of metals into MIBK

Metal added, mcg		Metal found, mcg	Recovery of added metal, mcg	% Recovery
	0.0	0.0		
	0.0	0.0		
	0.0	0.0		
	2.0	1.7	1.7	85
Cd	2.0	1.8	1.8	90
	4.0	3.7	3.7	93
	4.0	3.5	3.5	88
	4.0	4.0	4.0	100
				———
				91 mean
	0.0	3.4		
	0.0	3.2		
	0.0	3.5		
Cu	2.0	5.7	2.3	115
	4.0	7.6	4.2	105
	4.0	7.6	4.2	105
				———
				108 mean
	0.0	0.5		
	0.0	0.5		
Zn	1.0	1.5	1.0	100
	2.0	2.4	1.9	95
	3.0	3.4	2.9	97
				———
				97 mean

[a] 100–ml. volumes.

CONCLUSIONS

Pyridine-2-aldehyde-2-quinolylhydrazone has been shown to be a useful semiselective scavenging reagent for use in atomic absorption spectroscopy. Because it complexes a relatively small number of metals, it may find use in situations where large quantities of a particular metal would lead to excessive consumption of a less selective reagent such as ammonium pyrrolidinedithio-carbamate (APDC) or dithizone. PAQH is also more stable in solution than these two reagents.

The procedure described in this paper was developed for a 100-ml sample and gives a concentration factor of 12.5 (since the solubility of MIBK in water is about 2%). Slight modifications should allow the procedure to be applied to larger sample volumes.

Acknowledgements

This work was supported by grants from the National Research Council and the Department of Energy, Mines, and Resources of Canada.

References

1. M. Heit and D. Ryan, *Anal. Chim. Acta* **34**, 307 (1966).
2. R. E. Jensen and R. T. Pflaum, *Anal. Chim. Acta* **37**, 397 (1967).
3. S. P. Singhal and D. E. Ryan, *Anal. Chim. Acta* **37**, 91 (1967).
4. B. K. Afghan and D. E. Ryan, *Anal. Chim. Acta* **41**, 167 (1968).
5. G. G. Sims and D. E. Ryan, *Anal. Chim. Acta* **44**, 139 (1969).
6. R. E. Jensen, N. C. Bergman, and R. J. Helvig, *Anal. Chem.* **40**, 624 (1968).
7. V. Zatka, J. Abraham, J. Holzbecher, and D. E. Ryan, *Anal. Chim. Acta*, in press.
8. R. W. Frei, G. H. Jamro, and O. Navratil, *Anal. Chim. Acta*, in press.
9. M. Heit and D. E. Ryan, *Anal. Chim. Acta* **32**, 448 (1965).
10. J. F. Geldard and F. Lions, *Inorg. Chem.* **2**, 270 (1963).

The Determination of Trace Transition Elements in Biological Tissues Using Flameless Atom Reservoir Atomic Absorption[†]

D. A. SEGAR and J. L. GILIO

Rosenstiel School of Marine and Atmospheric Science, University of Miami, 10 Rickenbacker Causeway, Miami, Florida 33149, U.S.A.

KEY WORDS: Biological tissues, trace metals, flameless atomic absorption, heated graphite atomizer.

The development of an analytical technique generally applicable to the determination of Ag, Cd, Co, Cu, Fe, Mn, Ni, Pb, V, and Zn in all biological tissues is described. All of these elements may readily be determined in tissue samples of less than 0.5 g using flameless atom reservoir atomic absorption with a Perkin-Elmer heated graphite atomizer. Some of the operational characteristics of this atomizer system are discussed. The utility of the method is illustrated by analyses of selected marine biota and of NBS standard orchard leaves.

INTRODUCTION

The analysis of biological tissues for their concentrations of trace transition elements has long been a problem of interest to environmental scientists.[1] Many techniques, such as absorption spectrophotometry, polarography, emission spectrography, gas liquid chromatography, flame atomic emission and absorption, and neutron activation have been employed for such analyses.[1-3] Most of the analytical procedures that have been developed utilizing these techniques have been concerned with the analysis of a single element or at most two or three elements. However, it is desirable to determine

---

† Contribution Number 1628 from the Rosenstiel School of Marine and Atmospheric Science.

a number of elements simultaneously in the same sample, and this has been achieved by the application of atomic absorption,[4–8] optical spectrography[9–14] and neutron activation analysis.[15–16]

The flame atomic absorption determination method has been extensively utilized in trace transition element analysis of biological tissues.[4–8] Atomic absorption is rapid, simple, relatively free from interferences, and inexpensive in equipment requirements as compared to most other techniques. However, its application has been limited because the attainable sensitivities both in terms of concentration and absolute mass are somewhat poorer than those that may be obtained with, for example, neutron activation analysis.

The recently developed flameless atom reservoir atomic absorption technique has improved the attainable sensitivity of atomic absorption spectrophotometry by at least an order of magnitude for most elements.[17,18] The increase in sensitivity is such that for many elements this technique offers detection limits at least as good as, and often better than those attainable with any other currently available elemental analysis method. Thus, the flameless atom reservoir atomic absorption technique appears to be ideally suited to the development of a comprehensive analytical method for tissue samples. In this paper we describe the development of such a method for determination of a range of transition elements, and its application to tissue samples of marine origin.

EXPERIMENTAL

Reagents and apparatus

A Perkin-Elmer Model 403 atomic absorption spectrophotometer equipped with a Sargent-Welch Model SRG recorder, a deuterium arc background corrector and a Perkin-Elmer HGA-70 heated graphite atomizer were employed.

The heated graphite atomizer electronics were modified to enable selective volatilization analysis to be carried out.[19] A modification was also employed which allowed the use of the static method of analysis to improve the sensitivity.[20] In our system, the solenoid valve is activated automatically by a relay at the commencement of the atomization cycle and a secondary timer-controlled relay allows the gas flow to be stopped for between 0 and 30 sec, independently of the atomization time. This replaces the manual operation of the solenoid employed earlier.[20]

All temperature settings used in the graphite atomizer are reported in terms of the applied voltage measured directly across the atomizer terminals, and the approximate corresponding temperature determined from the instrument manual is indicated in parentheses (Table I).

All sample injections were made with Eppendorf microliter pipettes with disposable plastic tips.

Nitric acid and methylisobutyl ketone and glacial acetic acid were purified by distillation from a silica still, and ammonia solution was prepared by dissolution of the gas in double-silica-distilled, de-ionized water. Ammonium pyrrolidine dithiocarbamate was recrystallized from alcohol-water, and its solution was extracted with several portions of ketone before use. Containment vessels and extracting funnels throughout the analysis were of fused quartz, Teflon, or high-density polyethylene.

Sample preparation and dissolution

All tissue samples were freeze-dried to constant weight before analysis. Until the recent development of a solid sample atomization technique,[21] each of the commercially available atom reservoirs has required that the sample be in solution for introduction into the reservoir. The development of the method described here was, therefore, based upon the use of dissolved samples for the analysis. Although it would seem that the necessity for utilizing dissolved samples may be obviated by the direct use of solid samples, our preliminary experience with solid biological tissue samples leads us to believe that this may not be true, except in limited instances. The results of our study of solid sample analysis will be reported in greater detail elsewhere. However, it appears that difficulties such as the long drying and ashing time required, matrix interferences, the lack of dynamic range, and nonreproducible volatilization of the non-uniformly dispersed samples may reduce the effectiveness of this technique. It would therefore appear that for general analysis of varied samples with a requirement for high accuracy and precision, the dissolution of the sample will still be necessary.

Many methods for the ashing and dissolution of biological tissues have been employed.[22,23] During analysis for trace transition metals, the major problem has often been that of either contamination of the sample with impurities in the reagents used or the loss of metal from the sample during ashing. To minimize both of these possibilities, we have chosen to wet-ash the organic matrix with nitric acid alone in pure fused silica conical flasks.[4] The flasks are heated on a hotplate and each flask is covered with a silica bulb stopper to maintain the acid vapor within the flask as long as possible. Contamination of the samples by this procedure is extremely small as the nitric acid may be obtained in a highly purified state by distillation from a silica still. Oxidation is somewhat slow when using nitric acid alone. However, we have always been able to achieve complete dissolution of the samples if the digestion is prolonged for a period of several days, except when small quantities of silica residues were left from some siliceous organisms. The possible loss of metal during the ashing was investigated using standard additions or

TABLE I

Recommended settings for Perkin–Elmer HGA-70 heated graphite atomizer[a]

	Sequence							Gas delay (sec)
	Drying[b]		Ashing		Volatilizing			
	Time (sec)	Voltage[c]	Time (sec)	Voltage[c]	Time (sec)	Voltage[c]		
Copper	30	0.4 ($\sim$ 100°C)	15	3.0 ($\sim$ 900°C)	20	7.0 ($\sim$ 2100°C)		4
Iron	30	0.4 ($\sim$ 100°C)	15	3.6 ($\sim$ 1150°C)	25	8.0 ($\sim$ 2200°C)		6
Zinc	30	0.4 ($\sim$ 100°C)	15	2.0 ($\sim$ 300°C)	15	7.5 ($\sim$ 2150°C)		2
Cadmium	30	0.4 ($\sim$ 100°C)	15	1.5 ($\sim$ 250°C)	15	6.5 ($\sim$ 1950°C)		2
Vanadium	30	0.4 ($\sim$ 100°C)	45	5.0 ($\sim$ 1300°C)	35	8.5 ($\sim$ 2300°C)		8
Nickel	30	0.4 ($\sim$ 100°C)	15	3.0 ($\sim$ 900°C)	25	8.0 ($\sim$ 2200°C)		6
Lead	30	0.4 ($\sim$ 100°C)	20	1.5 ($\sim$ 250°C)	20	7.0 ($\sim$ 2100°C)		4
Silver	30	0.4 ($\sim$ 100°C)	15	2.0 ($\sim$ 300°C)	20	7.5 ($\sim$ 2150°C)		4
Manganese	30	0.4 ($\sim$ 100°C)	20	2.0 ($\sim$ 300°C)	20	7.0 ($\sim$ 2100°C)		4
Cobalt	30	0.4 ($\sim$ 100°C)	15	3.6 ($\sim$ 1150°C)	20	8.0 ($\sim$ 2200°C)		6

[a] For determination of aqueous 0.1N HNO_3 solution of metals.

[b] Drying times are for 20-mcl injections. For smaller volumes, drying time may be decreased. For larger volumes, drying time should be increased (e.g., 100-mcl injection, 60 sec drying).

[c] Voltage as read between + and − terminals of the atomizer.

radiochemical tracers. No loss of Zn, Pb, Co, Ag, Sn, Fe, Mn, Cr, Cd, Ni, or Cu was observed.

Analysis of the aqueous digest solution by direct injection into the heated graphite atomizer.

It would be ideal if aliquots of the aqueous solution of the ashed biological tissue could be used directly for injection and analysis with the heated graphite atomizer. However, our experience has been that this is not possible for most samples when determining trace transition elements. The complicated matrix composition of the solution does not normally permit accurate determinations to be carried out by this means. The concentrations of most transition elements in biological tissues are almost always very small compared to the concentrations of the major metallic elements, particularly sodium, potassium, calcium, and magnesium. Thus, an aliquot of the ashed solution of most samples which contains sufficient transition metal for analysis by atom reservoir atomic absorption also contains relatively large quantities of other inorganic salts. These salts are atomized at the same time as the transition elements and due to their mass they give rise to considerable non-specific molecular and scattering absorption which interferes with the determination. A small amount of such interference can be compensated for by the use of a deuterium arc background corrector, but in practice the magnitude of the interfering absorption was found to be too great in many analyses and the correction was no longer effective.

For those elements that are relatively involatile, selective volatilization analysis is possible, whereby the major interfering salts are removed at a low temperature and then the element of interest is atomized at a higher temperature.[19,24] However, there is almost always some co-volatilization of the analysis element during the volatilization of the major salts. The quantity of the analysis element co-volatilized was found to be critically dependent upon the total salt content of the solution[24] and also upon the major element, both cation and anion, composition of the dissolved salt matrix. As both the total salt content and the matrix composition may be extremely variable in the ashed solutions of biological tissues, the selective volatilization technique is not applicable, except in instances where a routine analysis is to be performed for a series of tissue samples of identical compositional types. In addition, the large concentrations of calcium often present in biological tissues make the major salt matrix more refractory than was the case in a previous investigation of sea water analysis by this technique.[24] Thus, it was necessary to raise the ashing temperature to effectively remove the scattering interference and consequently co-volatilization of the analysis element was increased and the sensitivity reduced.

In the light of the above difficulties, it was found necessary to extract the transition elements from the major elements before determination. This extraction also provides an increase in the overall sensitivity of the analysis. The ash from as little as 1 g of biological tissue, depending upon the actual tissue type, often requires at least 20 ml of nitric acid for complete dissolution. After extraction, this volume can be reduced, and a larger proportion of the sample may be introduced into the atomizer for each analysis.

Solvent extraction of transition metals

The solvent extraction of transition elements as their pyrrolidine dithio-carbamates has been extensively utilized in the separation of these elements from the alkali and alkaline earth metals prior to analysis by flame atomic absorption.[4] This extraction technique was adopted for the current study, as it had already been shown to be quantitative for Zn, Pb, Co, Ni, Cd, Fe, Mn, Sn, Cu, and Ag extracted into methylisobutyl ketone or n-amylmethyl ketone from a digest solution of various marine animal tissues.[4] In the present work, we have also found the ashing and extraction technique to be quantitative for vanadium.

The ketone solution of the metal pyrrolidine dithiocarbamates may be injected into the heated graphite atomizer for analysis of the various metals.

TABLE II

Approximate detection limits using the HGA-70[a]

	Approximate detection limits (ng/g)	
	Analysis of ketone extract (5-mcl injections)	Analysis of extract after reversion to aqueous solution (100-mcl injections)
Ag	1.2	0.02
Cd	1.8	0.03
Co	30	0.5
Cu	30	0.5
Fe	25	0.4
Mn	25	0.4
Ni	120	2.0
Pb	20	0.3
V	300	5.0
Zn	0.6	0.01

[a] 1 g sample extracted into 15 ml ketone, reverted to 5 ml of aqueous solution.

However, the detection limits are not significantly better than those that could be obtained simply by aspirating the ketone directly into the flame atomizer, as we have done in previous investigations[4] (Table II). This lack of sensitivity is due to the necessity for restricting the volume of ketone injected to 5 mcl or less. Due to the low surface tension of organic solvents as compared to water, ketone, when injected into the atomizer, spreads out on the floor of the tube. If more than 5 mcl are injected, some of the ketone is lost from the ends of the atomizer and the analysis is not reproducible. Multiple additions of 5 mcl of ketone at a time, with drying between injections, can be used to increase the total sample weight introduced into the atomizer. However, the reproducibility of analysis was much reduced when using multiple injections. Therefore, it was decided that the transition metals should be reverted back to aqueous solution before analysis. Currently under assessment are modified graphite tubes which eliminate loss of organic solvents during drying. When these tubes become more freely available, reversion back to aqueous solution should prove unnecessary for most analyses. Modified tubes were not, however, available for this investigation.

Redissolution in aqueous solution

The most rapid means of returning trace metals in chelate form from organic solution to aqueous solution is by back extraction, usually with an acid solution. However, the nature of the equilibrium between aqueous and organic solvents of the metal pyrrolidine dithiocarbamates is such that even when strong acids were used for back extraction (e.g., 6N HNO_3), a proportion of some metals, e.g., Pb, Cd, V, remained in the organic phase. Therefore, the metals were returned to the aqueous phase by evaporating off the ketone solvent, destroying the organic components of the resultant matrix with nitric acid and dissolving the metals in dilute nitric acid solution. This procedure has been found to be quantitative and the resultant inorganic salts may be dissolved in as little as 5 ml volume or less.

Atomic absorption determination

The range of trace transition elements to be found in biological tissues is extremely large. Therefore, it was found advantageous to utilize the very large dynamic range offered by the parallel use of flame atomization and the heated graphite atomizer (Table III). Flame atomization atomic absorption can be carried out about twice as fast as flameless atom reservoir atomic absorption, so that the samples were initially monitored by this technique to determine whether or not concentrations of the various elements were below the optimum concentration range. For those elements whose concentrations were within the optimum range for flame atomization analysis or were above this range (normally only Fe and Zn), analysis was carried out using standard flame atomic absorption conditions[25] and either dilutions of, or the undiluted, aqueous solutions. For those elements whose concentrations were below this range, normally Ni, Co, Cd, Pb, Ag, V and for most of our samples Cu, the undiluted aqueous solution or dilutions of these were utilized for analysis by injection into the heated graphite atomizer.

TABLE III

Optimum concentration ranges for flame and heated graphite atomizer analysis

	Concentration range (ng/ml) in analysis solution		ng/g Sample[a]	
	Flame	HGA–70	Flame	HGA–70
Ag	1,000– 10,000	0.004– 500	5,000– 50,000	0.02– 2,500
Cd	500– 5,000	0.006– 500	2,500– 25,000	0.03– 2,500
Co	2,000– 20,000	0.1 –1,000	10,000–100,000	0.5 – 5,000
Cu	2,000– 20,000	0.1 –3,500	10,000–100,000	0.5 –17,000
Fe	2,000– 20,000	0.1 –1,000	10,000–100,000	0.4 – 5,000
Mn	1,000– 10,000	0.1 – 400	5,000– 50,000	0.4 – 2,000
Ni	2,000– 25,000	0.5 –4,000	10,000–125,000	2.0 –20,000
Pb	4,000– 40,000	0.1 –2,000	20,000–200,000	0.3 –10,000
V	5,000–100,000	1.0 –5,000	25,000–500,000	5.0 –25,000
Zn	200– 3,000	0.002– 200	1,000– 15,000	0.01– 1,000

a 1 g sample dissolved in final volume of 5 ml.

Injection volumes of from 5 mcl to 100 mcl were employed as necessary. The optimum conditions for analysis utilizing the heated graphite atomizer have been determined (Table I). For each metal, an ashing step was employed during which the furnace was heated to a temperature just insufficient to atomize any of the analysis element. For most samples, this step was unnecessary. However, the step was always utilized as a precaution to remove any volatile material and thus minimize the scattering signal during the analysis atomization. When using these conditions combined with the use of the deuterium arc background corrector, no interferences have been observed for a considerable range of samples. Chemical interferences were also absent as determined by the analysis of standard solutions of each analysis element in the presence of large excesses (5,000 fold) of each of the other elements determined.

Adopted procedure

The freeze-dried tissue samples were ashed as previously described.[4] Samples of ca. 1 g dry weight in 100-ml fused silica conical flasks were used.

The solution of the ashed tissue sample was diluted to 100 ml, transferred to a 250-ml Teflon separatory funnel, 1 ml of glacial acetic acid was added and the pH was adjusted to ca. 3.0 by cautious addition of ammonia solution. A 10% aqueous solution of ammonium pyrrolidine dithiocarbamate (8 ml) was added. After mixing thoroughly, the pH of the solution was adjusted to 3.0–3.5 and the transition metals were extracted immediately into 15 ml of methylisobutyl ketone. After the two layers had separated, the organic phase was transferred quantitatively to a 100-ml fused silica conical flask. The aqueous solution was re-extracted with another 10 ml of methylisobutyl ketone and the organic phase was combined with that from the first extract. The silica flask was heated on a hotplate at ca. 100°C to evaporate the ketone. When the ketone had been completely evaporated, the resultant organic matrix was ashed with 5 ml of concentrated HNO_3. The solution was allowed to evaporate almost to dryness and 0.1 N nitric acid was added to make the volume up to about 3 ml. The solution was transferred quantitatively to a 5-ml graduated flask, made up to volume with 0.1 N nitric acid and transferred to a 10-ml polyethylene bottle for storage prior to atomic absorption analysis.

Analysis was carried out either by flame atomization of this solution or dilutions of it, or by flameless atom reservoir atomic absorption spectrophotometry utilizing injection volumes of 5–100 mcl and at least three replicate injections (see Tables I and III).

The atomic absorption spectrophotometer was calibrated with aqueous standards prepared from ultra-pure chemicals (Alfa Inorganic minimum 99% pure). Corresponding blank determinations and standards carried throughout the ashing, extraction and analysis procedure were utilized. The precision of the method depends upon the concentration of the element to be determined but in no instance was it worse than $\pm 10\%$ (standard deviation 3 replicates) except when samples had concentrations very close to the detection limits (signal/noise ratio of less than 10/1).

RESULTS AND DISCUSSION

The technique described above has been applied to the analysis of many samples of marine grasses, algae and sponges from a sub-tropical estuary. These samples represent a considerable spectrum of different tissue types particularly with regard to the degree of calcification. The results of these analyses will be reported elsewhere. However, examples of the results obtained are reproduced in Table IV. In addition, samples of NBS standard orchard leaves (SRM–1571) have been analyzed for some elements (Table V). It may be noted that this standard reference material contains a fraction considerably more resistant to oxidation than we have found in marine biota samples

TABLE IV

Selected analyses of marine biota[a] (mcg/g dry weight)

	Ag	Cd	Co	Cu	Fe	Mn	Ni	Pb	V	Zn
Marine grass										
Thalassia										
testudinum	0.15	0.6	0.94	3.72	64.4	81.4	6.6	1.65	6.15	23.2
Marine algae										
Laurencia										
poitei	—	0.5	0.13	3.70	175	50.3	1.5	0.86	4.94	15.4
Penicillus										
capitatus	—	0.4	0.05	0.86	81.1	33.4	0.6	1.18	7.03	8.57
Marine sponges										
Chondrilla										
nucula	0.03	2.0	0.87	15.1	215	9.67	8.5	0.87	13.5	50.0
Haliclona										
molitba	0.32	1.2	1.53	5.88	444	5.54	1.7	1.14	12.0	22.1

[a] All results are based on the average of three or more sample injections.

TABLE V

Analysis of NBS standard orchard leaves SRM 1571

Element	Concentration determined[a] (mcg/g)	Certified concentration (mcg/g)	
Fe	290	300	± 20
Mn	87	91	± 4
Pb	44	45	± 3
Zn	24	25	± 3
Cu	12	12	± 1
Ni	1.4	1.3	± 0.2
Cd	0.23 ± 0.02	0.11	± 0.02

[a] Average of at least four samples. Cd value is average of 10 samples.

from all but a very few species. It was, therefore, necessary in addition to the nitric acid to use a small quantity (1 ml) of perchloric acid to complete the oxidation, and hydrofluoric acid (2 ml) to dissolve the silica residue.

Analysis of these samples has shown the applicability of the current method to all biological samples and the method may be used to great effect when it is required to analyze a range of different samples. For analysis of samples of similar composition, more direct analytical procedures may be applicable, particularly when analysis for the less volatile elements only is desired. Development of such specific methods is currently being carried out particularly with reference to blood and urine samples.[26]

Acknowledgements

This work was supported by the National Science Foundation, Grants GA 33003 and GU 103302 Sub 27, AEC contract AT-40 (1) 3801, Sub 2, and NOAA Sea Grant 2-35147.

References

1. H. J. M. Bowen, *Trace Elements in Biochemistry* (Academic Press, London, 1966).
2. A. P. Vinogradov, *The Elementary Composition of Marine Organisms* (Sears Found. Mar. Res., Yale Univ., Memoir No. II, 1953).
3. E. D. Goldberg, *Review of trace element concentrations in marine organisms* (Puerto Rico Nuclear Centre, Puerto Rico, 1965).
4. J. P. Riley and D. A. Segar, *J. Mar. Biol. Ass. U.K.* **50,** 721 (1970).
5. J. H. Martin, *Limnol. Oceanogr.* **15,** 756 (1970).
6. J. H. Martin, *Bioscience* **19,** 898 (1969).
7. R. A. Stevenson, S. L. Ufret, and A. T. Diecidue, *Proc. 5th Intern. Amer. Symp. on the Peaceful Applications of Nuclear Energy, Pan Amer. Union.*, Wash. D.C., p. 233 (1965).
8. B. H. Pringle, D. E. Hissong, E. L. Katz, and S. T. Mulawaka, *J. Sanit. Eng. Div.* **94,** 455 (1968).
9. W. A. P. Black and R. L. Mitchell, *J. Mar. Biol. Ass. U.K.* **30,** 575 (1952).
10. I. Noddack and W. Noddack, *Ark. Zool.* **32A,** 1 (1940).
11. D. A. Webb and W. B. Fearton, *Sci. Proc. R. Dubl. Soc.* **21,** 487 (1937).
12. D. A. Webb, *Sci. Proc. R. Dubl. Soc.* **21,** 505 (1937).
13. E. G. Young and W. M. Langille, *Can. J. Botany* **36,** 301 (1958).
14. J. P. Riley and I. Roth, *J. Mar. Biol. Ass. U.K.* **51,** 63 (1971).
15. H. F. Lucas, Jr., D. N. Edington, and P. J. Colby. *J. Fish. Res. Bd. Can.* **27,** 677 (1970).
16. H. D. Livingstone and G. Thompson, *Limnol. Oceanogr.* **16,** 786 (1971).
17. B. V. L'Vov, *Atomic Absorption Spectrochemical Analysis* (Elsevier, New York, 1970).
18. D. C. Manning and F. Fernandez, *At. Absorption Newslett.* **9,** 65 (1970).
19. D. A. Segar and J. G. Gonzalez, *At. Absorption Newslett.* **10,** 94 (1971).
20. H. L. Kahn and S. Slavin, *At. Absorption Newslett.* **10,** 125 (1971).
21. J. D. Kerber, *At. Absorption Newslett.* **10,** 104 (1971).
22. G. Middleton and R. E. Stuckey, *Analyst* **78,** 532 (1953).
23. Analytical Methods Committee, *Analyst* **85,** 643 (1960).
24. D. A. Segar and J. G. Gonzalez, *Anal. Chim. Acta* **58,** 7 (1972).
25. Perkin Elmer Corp., *Analytical Methods for Atomic Absorption Spectrophotometry*, Perkin Elmer Corp., Conn. (1971).
26. R. T. Ross, J. G. Gonzalez, and D. A. Segar, *Anal. Chim. Acta*, **63,** 205 (1973).

Activation Analysis and Applications to Environmental Research

R. H. FILBY and K. R. SHAH

Department of Chemistry and Nuclear Radiation Center
Washington State University Pullman, Washington 99163

The principles of activation analysis as a method of trace element determination are discussed and applications of neutron activation analysis to environmental problems reviewed. Thermal and fast neutron activation analysis using nuclear reactors, accelerators and ^{252}Cf sources are considered and recent developments in γ-ray spectrometry such as Ge(Li) detectors, anticoincidence shielded Ge(Li) detectors and multidimensional NaI(Tl) spectrometers are outlined. The advantages and disadvantages and sources of error in neutron activation analysis for trace element determination are discussed with emphasis on problems associated with the analysis of environmental materials. Applications of neutron activation analysis to air pollution studies, marine and fresh-water trace element measurements and to the problem of mercury pollution are reviewed.

Contents

INTRODUCTION

Concern about the effects of toxic elements in the environment and as yet little-known effects of other elements introduced into the environment has resulted in increased interest in the measurement of such elements in environmental materials. Gross contamination of the environment is usually easy to detect but many scientists are now aware that even the introduction of minute amounts of certain elements, for example, mercury and cadmium, may have deleterious effects if continued over long periods of time. The concentrations of many elements and their compounds in the undisturbed environment (i.e. "background" concentrations) are not known and their geochemical behaviour in the hydrosphere and atmosphere is not well established. In particular, very little is known of the behaviour of various chemical forms of an element in natural processes and their distribution in aquatic food chains. The establishment of water- and air-quality standards is complicated both by lack of knowledge of "natural" or "background" concentrations and the difficulties of analytical determination caused by the very low concentrations of many elements in environmental materials. Thus considerable work remains to be done on the measurement of very low elemental concentrations and the distribution of elements in the environment. Animal nutritional studies have also shown the importance of excesses or deficiencies of certain trace elements in soils and associated forage crops and trace element levels in soils and plants have received considerable attention.

The scope of this article is to outline the principles of activation analysis and to review some applications of the technique to the determination of trace elements, including toxic elements, in environmental materials. Whether or not an element may be regarded as toxic depends on the organism affected, the degree of toxicity, the type of toxicity, and the chemical state of the element. In this review we have included those elements known, or suspected, to be toxic either to man or to ecologically important organisms. We have also included in the discussion other trace elements which are not regarded as toxic in the forms and amounts encountered in environmental materials. For many of these elements, however, insufficient information is available on their ecological effects and other elements may have no direct harmful effects yet may cause environmental damage through secondary effects. An example of the latter is the stimulation of rapid algal growths (blooms) by elements such as phosphorus or molybdenum. The environmental materials considered in this review comprise water, air, air particulates, and biological materials if related to environmental problems. Table 1 lists some of the more

TABLE I

Nature and source of pollution of some elements of environmental concern

Element	Nature of pollution	Possible pollution sources
P	water pollutant	sewage; fertilizer run-off
S	air pollutant	fossil-fuel burning (coal, air)
V	air pollutant	residual fuel-oil burning
Ni	air pollutant	residual fuel-oil burning
Cu	water pollutant; air particulates	industrial discharges
Zn	water pollutant; air particulates	mineral smelting; automobile tires
Ge	air pollutant	coal burning
As	water pollutant; soil pollutant	insecticides; herbicides; phosphates
Se	air particulates	fossil-fuel burning
Br	aerosols and air particulates	automotive exhausts
Mo	water pollutant	industrial wastes; fertilizers
Cd	water pollutant; air particulates	industrial wastes; automobile tires
Sb	water pollutant; air particulates	industrial wastes; mineral processing
Hg	water pollutant; air particulates	chlor-alkali and pulp mills; fungicides
Pb	air pollutant; water pollutant	gasolines; paints; mineral processing

important elements of environmental concern, the nature of the problems with which they are associated and possible sources of these elements.

The determination of trace elements in environmental materials is complicated by the low elemental concentrations and the chemical complexity of the materials under investigation. Water, for example, may contain elements in soluble forms (anionic, cationic and neutral complexes) as well as in colloidal or particulate forms. Slight changes in conditions, for example pH, may affect the distribution of an element among the different states. A satisfactory analytical method for the determination of a given element should satisfy the following criteria:

a) high sensitivity, often better than 1 ng;

b) good precision and accuracy;

   c) high selectivity in the presence of other elements and freedom from
      interferences;
   d) be applicable to materials of a wide range of chemical compositions.

In addition to these criteria, other desirable properties of an analytical
method are:

   a) multielement technique;
   b) simple, rapid and inexpensive;
   c) identifies chemical form.

Most analytical methods currently in use for the determination of trace
elements in environmental materials satisfy the above criteria, at least for
some elements. The physicochemical methods include flame and flameless
atomic absorption spectroscopy, atomic fluorescence, emission spectrography,
mass spectrometry, colorimetry, X-ray fluorescence analysis, polarography,
anodic stripping voltammetry, and other electrochemical methods. All of
these techniques possess certain advantages and disadvantages, depending
on the element to be determined, the material to be analyzed, and the sensitivity required.

Activation analysis is a nuclear method of analysis which has developed
rapidly during the last twenty years and is now one of the most useful methods
of trace analysis. The technique meets all of the analytical criteria for many
elements of the periodic table and the sensitivities attained are often orders
of magnitude higher than for other methods. Recent developments in gamma-
ray spectrometry have made multielement analysis possible in a wide variety
of materials. Most of the early applications of activation analysis were in
materials science and in geochemical, cosmochemical or biological research,
but in the past decade activation analysis has been applied to a variety of
environmental problems and the usefulness of the technique should increase
in the future as instrumentation improves. The advantages and disadvantages
of activation analysis in environmental research will be described in the
following sections, but it should be pointed out that the technique does not
compare unfavorably in cost with many other methods. A complete multielement analysis system, excluding irradiation facilities, costs less than
$20,000. For many applications it is not necessary to possess a nuclear
reactor as more than sixty research and production reactors with irradiation
facilities available to outside users are to be found in the United States and
Canada alone. Many facilities perform irradiations at nominal costs, or at
no cost under Reactor Sharing Programmes. Portable neutron generators
($10,000–$20,000) are available in many institutions and the increasing
availability of intense ^{252}Cf neutron sources will increase the application of
activation analysis in many laboratories not having access to a nuclear
reactor.

PRINCIPLES OF NEUTRON ACTIVATION ANALYSIS

Several excellent recent texts and monographs (1–4) should be consulted for full account of activation analysis principles. The principle of activation analysis is the bombardment of an element with neutrons, charged particles, or photons thus causing a nuclear reaction to take place. The products of the reaction, if radioactive, may be identified by such specific nuclear properties as gamma-ray energies and intensities, beta-ray energies, or half-lives. The analytical procedure falls into two parts: irradiation of the sample and the measurement of the induced radionuclides. In radiochemical activation analysis, the nuclide of interest (or group of nuclides) is separated from other activities and measured, whereas in instrumental activation analysis the nuclide is measured without separation. Most activation-analysis methods for trace-element determination employ neutrons because cross sections for thermal-neutron induced reactions are generally much higher than those of charged-particle or photonuclear reactions. Also, high neutron fluxes are available in nuclear reactors and convenient portable neutron generators and ^{252}Cf sources are available for use where access to a reactor is not possible.

The reactions commonly used in neutron activation analysis (NAA) are the (n, γ), (n, α), $(n, 2n)$ and $(n, n'\gamma)$ and the reaction probability depends on the nature of the target nuclide and the neutron energy. The most important and widely used reaction in NAA is the (n, γ) capture reaction which takes place with thermal neutrons (0.025 eV most probable energy). All known stable nuclides, with the exception of ^{4}He, undergo the (n, γ) reaction and cross sections (i.e. reaction probabilities) for thermal neutrons are often very large (up to 240,000 barns). This fact, together with the fact that high thermal-neutron fluxes are available in nuclear reactors makes the (n, γ) reaction most generally applicable to activation analysis. Particle-emission reactions, (n, p), (n, α), $(n, 2n)$, and $(n, n'\gamma)$, have appreciable cross sections only at high neutron energies $(>1 \text{ MeV})$, because most reactions have positive threshold energies although a few light nuclei have high cross sections for (n, p) or (n, α) reactions with thermal neutrons. In most research reactors fast-neutron $(>1 \text{ MeV})$ fluxes are at least an order of magnitude lower than the thermal-neutron fluxes but the fast-neutron flux may be used for some analytical applications, for example, the determination of lead by the ^{204}Pb$(n, n'\gamma)^{204m}$Pb reaction (5). Neutron generators that produce high energy (usually 14 MeV) neutrons are generally used for fast neutron activation analysis.

For a nuclear reaction of the type $^A X(n, \gamma)^{A+1}X$, the induced activity A_t, in disintegrations/sec, at the end of irradiation of time t is given by:

$$A_t = N\sigma\phi(1 - e^{-\lambda t}) = N\sigma\phi(1 - e^{-0.693t/T}) \qquad (1)$$

where

 σ = thermal-neutron capture cross section in cm^2. The unit of cross section used is the barn (1 barn = 10^{-24} cm^2).

 ϕ = thermal neutron flux in neutrons cm^{-2}sec^{-1}.

 λ = decay constant of the product nuclide.

 T = half life of the product nuclide = $\dfrac{0.693}{\lambda}$

 N = number of atoms of $^A X$ in the target.

Figure 1 shows the variation of A_t as a function of irradiation time t. The

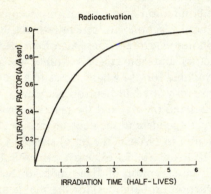

FIGURE 1 Growth of radionuclide activity (A/A$_{sat}$) as a function of irradiation time (in half lives).

expression for the activity, $A_{t'}$, of the product nuclide at any t' after the end of irradiation is given by:

$$A_{t'} = N\sigma\phi(1-e^{-\lambda t}).e^{-\lambda t'} \tag{2}$$

Consideration of Eq. (2) indicates that the following criteria should be met for a practical, sensitive neutron-activation method.

1) The product of the cross section, σ, and the isotopic fraction, f, of the parent nuclide should be as large as possible.

2) The half life of the product nuclide should be neither very long nor very short compared to the irradiation time.

3) The induced radionuclide should be readily detectable and measureable. For instrumental analysis using gamma rays the absolute gamma-ray intensities should be greater than 1% and the gamma-ray energies should be in the range 100–2000 keV.

The disintegration rate of the product nuclide, which is proportional to the weight of the element in the sample, may be determined by measurement of beta-ray or gamma-ray intensities. For instrumental analysis beta rays are not not used because the beta rays from a given nuclide are not monoenergetic and a chemical separation of the nuclide of interest must invariably be made. As most radionuclides emit gamma rays of characteristic energies, gamma-ray spectrometry is the preferred detection method for both radio-chemical and instrumental activation analysis.

Equation (2) indicates that, in principle, neutron activation is an absolute method of analysis. If t, t', σ, ϕ, λ, and A_t are known, or can be measured, then N, the number of atoms of the target nuclide (and hence the weight of the element present) can be calculated. In practice, absolute activation analysis is rarely employed because of uncertainties in ϕ and σ and the necessity of measuring absolute disintegration rates.

In practice, most neutron activation methods use the comparator technique. This involves irradiating a standard, which contains a known amount of the element of interest, with the sample. Thus, for sample and standard:

$$A_{sa} = N_{sa}\sigma\phi(1 - e^{-\lambda t})$$
$$A_{st} = N_{st}\sigma\phi(1 - e^{\lambda t})$$

where

A_{sa} = activity (disintegrations/sec) of the nuclide in the sample at end of irradiation.

A_{st} = activity (disintegrations/sec) of the nuclide in the standard at end of irradiation.

Hence

$$\frac{A_{sa}}{A_{st}} = \frac{N_{sa}}{N_{st}} = \frac{W_{sa}}{W_{st}} = \frac{C_{sa}}{C_{st}}$$

where

W_{sa} = weight of the element in the sample in μg.

W_{st} = weight of the element in the standard in μg.

C_{sa} = measured activity of nuclide in the sample (proportional to A_{sa}).

C_{st} = measured activity of nuclide in the standard (proportional to A_{st}).

Therefore

$$\text{concentration of element } (\mu g/g) = \frac{C_{sa} \cdot W_{st}}{C_{st} \cdot W} \tag{3}$$

where

W = weight of sample in grams.

INSTRUMENTATION FOR ACTIVATION ANALYSIS

Neutron sources

The most useful irradiation facilities are provided by nuclear reactors, although neutron accelerators are more useful for fast-neutron activation analysis and the increasing availability of ^{252}Cf sources permits portable neutron sources with available neutron fluxes approaching those of small nuclear reactors.

(i) *Nuclear reactors.*   Nuclear reactors are the most intense neutron sources available for activation analysis but have the disadvantage of not producing monoenergetic neutrons. Most published work on the application of activation analysis to environmental problems has utilized research reactors where high thermal-neutron fluxes are available. Nearly all research reactors use ^{235}U enriched uranium as the fissionable material or fuel. Many research reactors have the core immersed in water (or less commonly D_2O) which acts as a coolant, and a neutron moderator and provides flexibility of experimental facilities. Most reactors have a variety of experimental irradiation positions and also pneumatic tube access to the core for rapid transit of sample from the core to an experimental facility.

The neutrons emitted in the ^{235}U fission process range in energy from 0 to 25 MeV and the average fission neutron energy is 2.0 MeV. As thermal neutrons (most probable energy 0.025 eV) are employed in most activation methods, irradiation positions are generally placed in graphite reflectors or in the water region where significant neutron moderation has occurred. Typical values for neutron fluxes of different energy groups for the Washington State University research reactor (graphite reflector position) are shown in Table II. For most environmental applications of activation analysis the (n, γ) reaction with thermal neutrons is used and a high thermal-to-fast neutron flux ratio is desired to prevent fast-neutron interference. This is often achieved in water-moderated reactors by irradiating in the thermal column (highly moderated region) of the reactor but some loss in sensitivity is inevitable because of low total fluxes in the thermal column.

Table III lists nuclear data of some thermal neutron (n, γ) reactions of importance in environmental trace element analysis.

In thermal neutron activation analysis using (n, γ) reactions, the fast neutron component of the neutron spectrum can give rise to interfering (n, p) or (n, α) reactions. On the other hand, the fast neutron component can be used for fast reactions where thermal reactions are not suitable. Examples of some fast-neutron reactions of environmental interest are included in Table IV. Where fast neutron activation is carried out with reactor neutrons, the thermal neutron component must be eliminated or reduced to prevent

TABLE II

Neutron fluxes as a function of neutron energy in theWashington
State University research reactor

Neutron energy range	Neutron flux $(n.cm^{-2}sec^{-1})$
0–0.625 eV	5.0×10^{12}
0.625 eV–9.11 keV	8.2×10^{11}
9.12 keV–1.04 MeV	6.0×11^{11}
> 1.05 MeV	5.1×11^{11}

TABLE III

Nuclear data on nuclides used for measurement of elements of environmental
interest (6, 7)

Element	(n, γ) Nuclide[a] formed	$f\sigma_{th}$[b]	Predominant decay mode[c]	Most important gamma rays (keV)[d]
P	14d-^{32}P	0.19	β^-	None
S	87d-^{35}S	0.0092	β^-	None
	5.1m-^{37}S	2×10^{-5}	β^-	3105(94)
V	5.8m-^{52}V	4.9	β^-	1434(100)
Ni	2.6h-^{65}Ni	0.0017	β^-	1115(15), 1482(25)
Cu	13h-^{64}Cu	3.1	EC	511(β^+)
	5.1m-^{66}Cu	0.71	β^-	1039(8.9)
Zn	243d-^{65}Zn	0.23	EC	1116(52), 511(β^+)
	13.8h-^{69m}Zn	0.019	IT	439(100)
Ge	83m-^{75}Ge	0.18	β^-	265(12)
	11h-^{77}Ge	0.015	β^-	211(32), 264(64), 416(31)
As	26h-^{76}As	4.5	β^-	559(45), 657(6.2)
Se	120d-^{75}Se	0.26	EC	136(57), 265(60), 280(25)
	17s-^{77m}Se	2.0	IT	162(100)
	57m-^{81}Se	0.05	β^-	103(99)
Br	4.4h-^{80m}Br	1.5	IT	37(38)
	35h-^{82}Br	1.6	β^-	554(71), 619(44), 777(84)
Mo	67h-^{99m}Mo	0.12	β^-	140(^{99m}Tc), 181(7.5), 740(14)
Cd	54h-^{115}Cd	0.32	β^-	492(8), 528(28)
Sb	64h-^{122}Sb	3.5	β^-	564(63)
	60d-^{124}Sb	1.4	β^-	603(98), 723(11), 169(52)
Hg	65h-^{197}Hg	1.3	EC	78(20)
	47d-^{203}Hg	1.2	β^-	279(81)
Co	5.2y-^{60}Co	3.7	β^-	1333(100), 1173(100)
Pb	3.3h-^{209}Pb	0.008	β^-	None
	0.8s-^{207m}Pb	0.007	IT	1060

[a] Half-life information; s-seconds, m-minutes, h-hours, d-days and y-years.
[b] $f\sigma_{th}$-product of isotopic fraction of parent nuclide and thermal neutron (n, γ) cross section.
[c] Decay mode: EC-electron capture; IT-isomeric transition.
[d] Gamma rays per 100 disintigrations in parentheses.

competing (n, γ) reactions and to reduce the total γ-ray activity of the sample for instrumental analysis. Irradiations are generally performed in cadmium containers ($\sigma_{th} = 2460$ barns) or in ^{10}B enriched boron containers ($\sigma_{th} = 3.9 \times 10^3$ barns) which act on thermal neutron filters but which are essentially transparent to fast neutrons.

Nuclear reactors of the research type are expensive ($500,000–$2,000,000) to install and maintain but more than 60 reactors in the United States are available to outside users for nominal irradiation costs. Available neutron fluxes range up to 10^{15} n.cm^{-2}sec^{-1} depending on reactor type.

TABLE IV

Fast neutron reactions of environmental interest (8, 9)

Element	Reaction	Half-life of product	σ(mbarns) for 14 MeV neutrons	σ(mbarns) for reactor neutrons[a]
P	^{31}P$(n, \alpha)^{28}$Al	2.3m	150	1.4
S	^{32}S$(n, p)^{32}$P	14d	350	154
V	^{51}V$(n, p)^{51}$Ti	5.8m	27	—
Cr	^{52}Cr$(n, p)^{52}$V	3.8m	78	—
Fe	^{54}Fe$(n, p)^{54}$Mn	310d	373	56
	^{56}Fe$(n, p)^{56}$Mn	2.6h	103	0.44
Co	^{59}Co$(n, \alpha)^{56}$Mn	2.6h	39	0.14
Ni	^{58}Ni$(n, p)^{58}$Co	71d	237	400
Cu	^{63}Cu$(n, 2n)^{62}$Cu	9.8m	550	—
Se	^{77}Se$(n, n'\gamma)^{77m}$Se	18s	—	—
Ag	^{109}Ag$(n, 2n)^{108}$Ag	2.3m	700	—
Cd	^{111}Cd$(n, n'\gamma)^{111m}$Cd	49m	—	—
Au	^{197}Au$(n, n'\gamma)^{197m}$Au	7.5s	—	—
Tl	^{203}Tl$(n, p)^{203}$Hg	47d	30	0.002
Pb	^{204}Pb$(n, n'\gamma)^{204m}$Pb	67m	160	9.4

[a] Effective neutron fission-spectrum cross section.

(ii) *Accelerators as Neutron Sources.* Accelerators, which are sources of charged particles, may be used to produce monoenergetic neutrons through (p, n), (d, n), $(^3$He, $n)$ and (γ, n) reactions. The most important reactions of this nature occur with light elements and these are listed with pertinent nuclear data in Table V.

The most useful of the reactions listed in Table V is the ^{3}H$(d, n)^4$He reaction which is highly exoergic and produces neutrons of 14 MeV energy. This reaction forms the basis of most commercial neutron generators and with deuteron bombarded targets of ^{3}H diffused into a metal (usually Pd) neutron outputs of up to 10^{12} n.sec^{-1} can be obtained, with useful 14 MeV neutron fluxes of up to 5×10^{10} n.cm^{-2}sec^{-1}.

Generally cross sections for fast-neutron reactions (see Table IV) are much lower than for (n, γ) thermal reactions. The low cross sections and low fast-neutron fluxes ($< 10^{11}$ neutrons $cm^{-2}sec^{-1}$) available from generators limits their application to problems where: (a) no suitable (n, γ) product exists; (b) the lower (n, γ) contribution gives much lower background; and (c) light elements which have low (n, γ) cross sections are to be measured, e.g. oxygen, silicon, nitrogen and carbon.

(iii) Other neutron sources. Radioisotopic neutron sources of the Pu-Be(α, n) or ^{124}Sb-Be(γ, n) type do not have sufficient neutron output of trace element

TABLE V

Nuclear reactions used to produce neutrons in accelerators or generators

Reaction	Threshold energy (MeV)[a]	Q-value (MeV)	Minimum neutron energy (MeV)[b]
^{2}H$(d, n)^3$He	0	3.27	2.45
^{3}H$(d, n)^4$He	0	17.6	14.1
^{3}H$(p, n)^3$He	1.02	-0.77	0.64
^{9}Be$(\alpha, n)^{12}$C	0	5.71	5.27
^{9}Be$(d, n)^{10}$B	0	3.79	3.50
^{7}Li$(p, n)^7$Be	1.88	-1.65	0.30
^{12}C$(d, n)^{13}$N	0.33	-0.28	0.003

[a] Threshold energy in zero for Q positive (exoergic). The Coulomb barrier determines the particle energy.

[b] Neutron kinetic energy at threshold bombarding energy.

analysis. Recently, however, the U.S. Atomic Energy Commission has made ^{252}Cf sources available. Californium-252 with a half-life of 2.55 years decays by spontaneous fission (3%) and by α-particle emission (97%). The total neutron output is 2.34×10^{12} n.sec^{-1}gm^{-1} ^{252}Cf and with subcritical (U or Pu) multiplication neutron fluxes of the order of 10^{12} n.cm^{-2}sec^{-1} per gm ^{252}Cf are possible. At present ^{252}Cf sources of approximately 10 mg are available and effective thermal neutron fluxes are of the order of 10^8 n.cm^{-2}sec^{-1} which is much lower than most reactor thermal neutron fluxes. The ^{252}Cf source does have the advantage of portability and a number of specific environmental activation methods have been published using ^{252}Cf (10–12). As costs decrease (at present $10 per μg) and source strengths increase, mobile activation analysis facilities and remote activation analysis (13) will be possible—including the possibility of environmental sensors.

Measurement of radionuclides

Several radiation detectors can be used to measure the radionuclides produced by irradiation of the element of interest. Proportional counters, Geiger counters, and liquid scintillation techniques, although highly efficient for β-particles have low efficiencies for γ-rays of energy greater than 100 keV. As the β-particles emitted by a radionuclide show a continuous energy distribution from 0 to E_{max}, proportional and Geiger counters can only be used when the radionuclide of interest has been separated in a radiochemically pure state. Fortunately most radionuclides produced by (n, γ) reactions emit γ-rays of characteristic energies and γ-rays spectrometry forms the basis of most modern activation analysis techniques used in environmental research.

Gamma-ray spectroscopy became a useful tool in activation analysis when the scintillation NaI(Tl) detector was introduced in 1948 and many published activation analysis methods have utilized NaI(Tl) detectors for γ-ray spectroscopy. NaI(Tl) detectors can be made in large volumes, and hence, have high efficiencies for γ-ray measurement. Unfortunately the resolution† is usually not better than 5–10 per cent, and multielement analysis is best used for materials giving relatively few γ-rays, well separated in energy. More complex spectra, such as those commonly obtained with environmental samples, result in many overlapping peaks and spectrum analysis techniques using computers (14) must be employed to resolve individual radionuclide contributions.

The introduction of solid-state Ge(Li) γ-ray detectors has increased greatly the scope of non-destructive activation analysis. Although commercially available detectors are much lower in efficiency than NaI(Tl) detectors because of the small volumes at present available, the excellent resolution of Ge(Li) detectors (FWHM‡, 2–3 keV at 1333 keV) allows even extremely complex spectra to be resolved with very few peaks overlapping. At present the size, hence, efficiency, of Ge(Li) detectors is governed by Li-drifting limitations and volumes of high resolution (<3 keV) detectors are limited to less than 100 cm³. The recent production of detectors from intrinsic germanium which does not need Li compensating holds promise for much larger detectors, although the present state of the art is restricted to detectors of a few cubic centimeters in volume. Table VI compares incident full-energy peak efficiencies, resolution (FWHM), peak/Compton ratios, and costs of a $3'' \times 3''$ NaI(Tl) detector and a 65 cm³ Ge(Li) detector.

A comparison of spectra obtained by NaI(Tl) and Ge(Li) detectors from a neutron-irradiated lung tissue sample is shown in Figure 2. The excellent

---

†Resolution is defined as $(\Delta E/E)100$, where $\Delta E = $ full-width at half maximum for γ-ray peak of energy E.

‡FWHM: full-width at half maximum peak height.

TABLE VI

Comparison of a 3″ × 3″ NaI(Tl) detector and a 65 cm³ Ge(Li) detector

Detector	Efficiency at[a] 1333 keV	FWHM at[b] 1333 keV	Peak to Compton ratio	Cost ($)
3″ × 3″ NaI(Tl)	1.2×10^{-3}	100–200	2:1 to 4:1	$900–$1,000
65 cm³ Ge(Li)	1.7×10^{-4}	2.1	35:1	$13,000

[a] Absolute full-energy peak efficiency at 25 cm.
[b] FWHM - full width of peak at half maximum height.

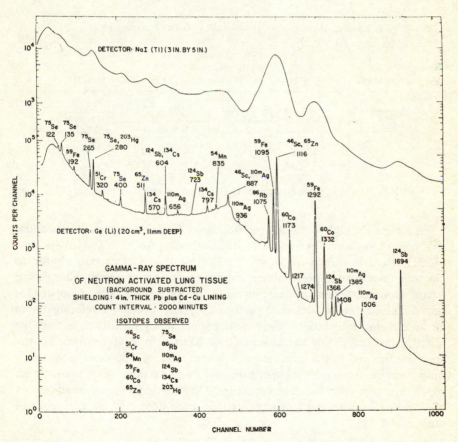

FIGURE 2   Gamma-ray spectra obtained by NaI(Tl) and Ge(Li) detectors of an irradiated lung tissue (from Cooper [15]).

resolution of the Ge(Li) detector allows the measurement of several radionuclides, e.g. ^{134}Cs, ^{54}Mn and ^{110m}Ag that cannot be measured in the NaI(Tl) spectrum without computer stripping routines. Also the NaI(Tl) detector does not resolve the important groups of gamma rays at 265–280 keV, 1075–1116 keV and 1292–1332 keV. The high resolution of Ge(Li) detectors not only permits more radionuclides to be detected and measured but also permits much simpler computer analysis programmes than are possible for NaI(Tl) spectrum analysis.

The γ-ray resolution of a Ge(Li) detector depends principally on the nature of the detector and the noise contribution of the associated elements (principally preamplifier noise).

ELECTRONICS BLOCK DIAGRAM OF BASIC Ge(Li) GAMMA-RAY SPECTROMETER

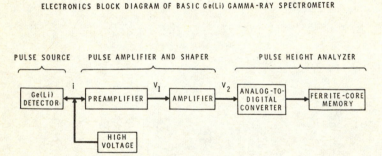

FIGURE 3   Schematic diagram of Ge(Li) spectrometer (from Cooper [15]).

To achieve the high resolution capabilities of Ge(Li) detectors, it is essential to employ low-noise electronics. A schematic diagram of a Ge(Li) spectrometer system is shown in Figure 3. The most important contributor to electronic noise is the preamplifier and improvements in preamplifier design permit close to theoretical detector resolution in certain cases. Cooper (15) has discussed Ge(Li) detector performance and recent developments in electronic instrumentation. The multichannel analyzer for collecting the spectral data commonly has at least 4096 channels for high resolution γ-ray spectrometry and possesses a range of output devices. The most satisfactory output devices are punched paper tape and magnetic tape which allow direct input to a computer for data reduction. Direct interfacing of analyzers or analog-digital converters to computers has been described by a number of authors (16, 17).

Compton suppression and coincidence techniques

It can be seen from Figure 2 that in a complex spectrum, the γ-ray peaks are superimposed on a Compton background caused by Compton scattering of γ-rays in the Ge(Li) detector followed by escape from the detector of the scattered photons. As the precision and sensitivity of analysis depend on peak area measurements, improved precision and sensitivity can be obtained by instrumentally suppressing the Compton background under the peaks. This is done by surrounding the Ge(Li) detector with a NaI(Tl) or plastic scintillator anticoincidence detector which cancels signals received by both detectors simultaneously, i.e. a Compton event. Figure 4 shows a schematic design of a Ge(Li)-NaI(Tl) coincidence-anticoincidence spectrometer (18) and Figure 5 shows normal and Compton suppressed spectra (19). Cooper has reported peak/Compton ratios of between 360 and 915 for a NaI(Tl)-Ge(Li) system compared to 35–40 for a good Ge(Li) detector without Compton

ANTICOINCIDENCE ELECTRONICS BLOCK DIAGRAM

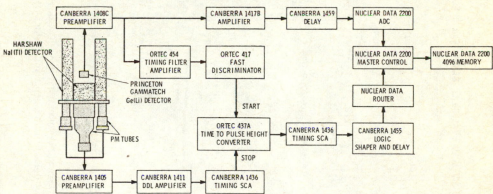

FIGURE 4   Schematic diagram of a Ge(Li)-NaI(Tl) anticoincidence spectrometer (from Cooper and Perkins [18]).

suppression. Cooper has also obtained high counting efficiency using two Ge(Li) detectors face to face surrounded by a plastic anticoincidence shield (19, 20).

The low efficiency of Ge(Li) detectors makes Ge(Li)-Ge(Li) coincidence spectroscopy useful only for intense sources, but Cooper (21) has described the advantages of such coincidence techniques applied to environmental problems. Ge(Li)-NaI(Tl) coincidence spectrometry using the NaI(Tl) pulse to gate the Ge(Li) spectrum has been proposed by a number of authors and can be applied to certain analytical problems of environmental interest (22, 23).

Large volume NaI(Tl) dual parameter spectrometers have found widespread use for low-level environmental radionuclide measurement, e.g. ^{60}Co, ^{137}Cs, etc., because of the very high detection efficiency (24, 25). Applications to

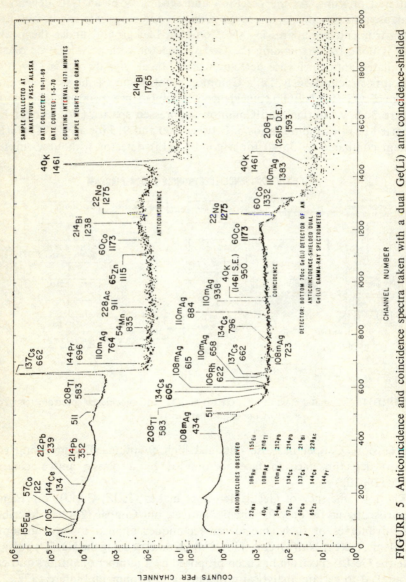

FIGURE 5   Anticoincidence and coincidence spectra taken with a dual Ge(Li) anti coincidence-shielded spectrometer (from Cooper and Perkins [19]).

activation analysis are limited but may be important for very low concentrations of elements whose (n, γ) capture nuclides possess coincidence gamma rays, e.g. ^{60}Co or ^{46}Sc.

ANALYTICAL PROCEDURES

Typical neutron activation analysis (NAA) procedures for environmental samples can be divided into a number of steps.

a) Sample preparation and preparation of elemental standards of known concentrations.

b) Irradiation of samples and standards in a neutron flux such that both receive the same neutron exposure.

c) Decay period to allow unwanted short-lived radionuclides to decay. This decay period will depend on the radionuclides to be measured.

d) Post-irradiation treatment of the sample. For instrumental neutron activation analysis (INAA) this step is usually confined to transfer of the irradiated material to a given counting geometry. Chemical separation of a radionuclide of interest may be necessary to improve sensitivity.

e) Measurement of radionuclide activity.

f) Computation of analytical results, usually performed by computer codes.

Sample preparation

One of the most important advantages of activation analysis is that usually no sample treatment is necessary prior to irradiation, hence contamination of the sample is avoided and the reagent blank is zero. In the preparation of environmental samples for irradiation, however, certain sources of error must be eliminated or reduced.

In many instrumental neutron activation analysis (INAA) methods, the sample must be packaged in a container and the sample plus the container is irradiated and then counted. It is important that the container material contain very low concentrations of the elements of interest. Table VII shows concentrations of some elements in several common container materials. For the determination of non-volatile elements, high-purity polyethylene is often satisfactory for high-flux and low-temperature irradiations. For volatile samples, or when elements may volatilize during irradiation (e.g. Hg), encapsulation in quartz is necessary.

The analysis of environmental and biological materials for trace elements presents some special sample-preparation problems. The analysis of water is perhaps the most difficult and for elements present at very low concentrations adsorption on suspended material or on the walls of the container may be significant. Robertson (27) has shown that serious losses of In, Sc, Fe, Ag, U and Co occur by adsorption from seawater onto polyethylene or glass containers over periods up to 90 days. For In and Fe up to 90 per cent adsorption was observed on polyethylene. Addition of hydrochloric acid to give a pH of 1.5 prevented adsorption on polyethylene except for Sc. For fresh waters adsorption of trace elements may be more serious than for seawater because of the lower total electrolyte concentration of fresh waters. For fresh waters acids are often added to prevent trace element precipitation and adsorption on container surfaces. The addition of acids to aqueous samples will however change the chemical nature of the medium. For example, certain trace elements may be present as suspended matter (e.g. silicates), finely divided precipitates of colloids (Fe compounds, Mn hydroxides, etc.), or bound to organic ligands such as humic acids. Addition of strong acids will change equilibria in the solution and change the concentrations of ionic species or

TABLE VII
Trace element contents of some irradiation container materials (26)

Material	Concentration (ppb)									
	Zn	Fe	Sb	Co	Cr	Sc	Cs	Ag	Cu	Hf
Polyethylene[a]	2.8	10^4	0.18	0.07	76	0.0008	0.005	0.1	6.6	0.5
Quartz tubing[b]	730	2.8×10^5	2.9×10^3	81	—	106	100	0.001	—	597

[a] Nalgene polyethylene.
[b] "Spectrosil" Thermal American Fuel Quartz Co.

complexes in true solution. The pre-irradiation treatment selected for water samples will depend on whether the total content of the element is required, or the distribution of the element among different phases in the sample. A combination of filtration, cation, and anion exchange has been proposed (28) as a method for distinguishing among particulates, cationic, anionic and uncharged states of trace elements in water. Several authors (29, 30) have frozen water samples immediately after sampling and in some reactors samples are irradiated frozen, thus preventing either adsorption of trace elements onto the surface of the container or desorption of impurities on the container surface. The problems encountered in analysis of water for trace elements are common to all analytical methods and have received insufficient attention to date. Although pre-irradiation processing of sample is generally avoided, certain problems may necessitate it. For example, it may be necessary to separate suspended material from an aqueous sample by centrifuging or filtering if both aqueous and suspended phases must be analyzed. Substantial nuclide activities may pass from the suspended to the aqueous phase by capture gamma-ray recoil following neutron capture thus making post-irradiation separation of phases subject to error. For extremely low trace element concentrations, preconcentration by evaporation of water samples or ion-exchange separation has been recommended by a number of authors (31, 32).

Air particulates or aerosols present few problems during irradiation. Such samples are generally obtained impregnated on a filter and the accuracy and sensitivity of the analysis will be affected by the composition of the filters used. Fiberglass filters are commonly used but contain high impurity levels and for low trace-element samples filter paper, polystyrene, Millipore, or Nucleopore (General Electric) filters are preferred.

Biological materials may be irradiated either in polyethylene containers or in quartz ampoules. However, Brune (23) has shown that if the irradiated material is washed out of the vial, contamination from the vial surface may result. Hence it is essential that high purity polyethylene or quartz vials be used for irradiation. It is often desirable to freeze-dry biological samples prior to activation to prevent high-pressure build up of water radiolysis products. Pillay *et al.* (34) have presented evidence that Hg is lost from biological samples during freeze-drying, perhaps as $Hg(CH_3)_2$ formed in the tissues. Experiments on ^{203}Hg-tagged tissues in this laboratory, however, indicated no losses of Hg during freeze-drying (35). Little information on the behaviour of other elements during freeze-drying is available. Further study of the freeze-drying process is underway in this laboratory.

The preparation of standards of accurately known composition is essential for precise analysis. Aqueous solutions of known elemental composition are commonly used as thermal neutron flux monitors for analysis of a wide variety of materials. Thermal neutron attenuation due to the major constituents of most environmental materials is similar to that in water and the nuclide specific activities obtained do not depend on chemical composition. For Ge(Li) spectrometry it is usually convenient to prepare multiple element solution standards, provided no chemical interactions among elements occur. For certain elements, care must be taken to ensure that the chemical form used is stable under the irradiation conditions used. For example, bromine as Br^- can form elemental Br_2 during irradiation (recoil effect) and a reducing agent is necessary to prevent this (36). Filby (36) used hydrazine sulphate for this purpose. For manganese, manganese(II) compounds are preferred to $KMnO_4$ which undergoes some reduction Mn(VII) $\rightarrow$ Mn(IV) resulting in MnO_2 precipitation during irradiation. The case of mercury is particularly interesting. Several authors (37–40) have noted that irradiation of Hg(II) solutions in polyethylene containers results in considerable ^{203}Hg loss by diffusion out of the system. The mechanism of this loss is under investigation in this laboratory and is possibly:

$Hg(II) \rightarrow Hg(I)$ reduction

$2Hg(I) \rightarrow Hg(II) + Hg(O)$ disproportionation

The Hg metal then diffuses out of the containers. An alternative explanation is that recoil ^{203}Hg atoms produced by ^{202}Hg neutron capture react with alkyl groups in the polyethylene to give volatile dialkyl mercury compounds. The problem is avoided in NAA methods by irradiation of samples and standards in quartz vials.

Although aqueous solution standards are extensively used in activation analysis, well calibrated natural materials are useful for specific applications. Several analyzed geochemical materials are available, the most useful of which are the U.S. Geological Survey standard rocks (G-1, G-2, W-1, etc.) which have been analyzed by a variety of techniques including activation analysis (41). Synthetic glasses of known trace element content are available from the National Bureau of Standards. Few biological standards are available but a standard kale (42) has been analyzed for many trace elements (43). The National Bureau of Standards has also issued a standard plant tissue (NBS 1571 Orchard Leaves) and is preparing standard blood and liver samples calibrated for contents of several trace elements.

Irradiation

When samples, together with corresponding standards, are irradiated in a reactor neutron flux it is important that all receive the same neutron exposure. In large graphite moderated reactors, flux gradients in irradiation positions are generally small but in most research reactors large thermal neutron flux gradients occur over short distances. In many reactors, groups of samples and standards are irradiated in a given position and are generally arranged in tiers, each tier containing a number of samples and standards. In the Washington State University research reactor graphite-reflector irradiation position the thermal neutron flux changes by a factor of 2.5 over a horizontal distance of 3.0 inches. Vertical flux gradients also exist, although less marked. To eliminate horizontal flux differences among samples on a given horizontal tier the samples are rotated about a vertical axis during irradiation thus equalizing the neutron exposure. Vertical differences among tiers or levels can be corrected for by irradiating standards on each tier, or by irradiating a flux monitor (Cu-wire, Au-foil, etc.) on each tier to obtain relative fluxes. This latter practice should be used with care because relative magnitudes of thermal and epithermal (n, γ) activation will be different for different nuclides. If fast-neutron reactions are being used, e.g. $^{58}Ni(n, p)^{58}Co$, standards of the element being measured must be used. When fast neutron reactions are used both sample and standard should contain equivalent amounts of light element (C, H, O, etc.) to eliminate neutron thermalization differences.

Accurate flux calibration and standardization is more difficult using 14-MeV neutron generators or ^{252}Cf sources where flux gradients are large.

Post-irradiation treatment

After irradiation and a suitable decay period, the radionuclide activities are determined using Geiger, proportional, NaI(Tl), or Ge(Li) detectors. For INAA using gamma-ray spectrometry no chemical treatment is performed on the sample. Samples and standards may be counted in the irradiation containers provided constant counting geometry is maintained but it is sometimes necessary to homogenize samples and to prepare both samples and standards in a non-radioactive constant matrix; e.g. as plaster-of-Paris or silica-gel discs. If samples are counted in the irradiation containers, the container activities must be accounted for. This can be done by irradiating blank containers.

If chemical separation of the radionuclides is necessary because of insufficient sensitivity by INAA or if the radionuclide is not a γ-emitter then a variety of chemical and physicochemical techniques can be used. The sample is first decomposed and brought into aqueous solution in the presence of known amounts of non-radioactive carriers of the elements of interest. The amounts chosen should be much greater than the amounts found in the sample. In addition, it is necessary that both the irradiated element in the sample and the elemental carrier be in the same chemical form prior to separation. For elements that have several stable oxidation states under the conditions employed, (Fe(III), Fe(II), Cl^-, ClO_4^-) it is necessary to oxidize or reduce both forms to achieve a common oxidation state. Separation procedures commonly include precipitation (44), solvent extraction (45, 46), complex formation (47, 48), ion-exchange (49, 50), distillation (51, 52) electrodeposition (53, 54), liquid-liquid extraction (55) and paper chromatography (56). The yield of the separation is measured by the amount of carrier recovered—thus quantitative separations are not required. This fact often allows highly specific, but not quantitative, separations to be made (e.g. many solvent extraction methods).

The high resolution of Ge(Li) detectors allows samples of complex mixtures of radionuclides to be measured, but often one or more nuclides present at high activities will mask most others. In environmental, biological and geochemical materials, the ^{24}Na activities of irradiated samples are often high enough to mask other nuclides of similar or shorter half-lives, e.g. ^{72}Ga, ^{42}K, ^{64}Cu. Removal of the interfering nuclide by a simple separation technique allows other nuclides present to be measured by Ge(Li) spectrometry.

Girardi and Sabbioni (57) have shown that hydrated antimony pentoxide (HAP) is highly selective for removing ^{24}Na from 8M H_2SO_4, only ^{182}Ta accompanies ^{24}Na. Other highly specific ion adsorbers have been developed, e.g., Al_2O_3, SnO_2, MnO_2, etc. (58).

Measurement of radionuclide activities

The irradiated material (or separated fractions) are generally counted for γ-ray emitting nuclides by NaI(Tl) or Ge(Li) gamma-ray spectrometry. Figure 6 shows a gamma-ray spectrum of irradiated blood and Figure 7 shows a spectrum of an irradiated crude oil. The parameter used as a function of nuclide activity is generally the area of a peak corresponding to a given gamma ray. As such peaks are superimposed on a continuous Compton background, peak areas must be obtained after background subtraction. For NaI(Tl) spectra the background may not vary in a linear fashion across the peak. A number of techniques (60–62) are used for the calculation of peak area, the most effective being that proposed by Covell (60).

For Ge(Li) spectra, the peaks occupy a relatively small portion of the energy range and except where peaks overlap or fall on a Compton edge, the background can be assumed to vary linearly from one side of the peak to the other. A number of authors have discussed techniques for the measurement of Ge(Li) spectra peak areas (63–65).

Digital computers have greatly assisted the analysis of complex gamma-ray spectra. With NaI(Tl) spectra analysis is usually performed by least squares fitting (66) or by stripping out normalized standard spectra sequentially (14) to correspond to the peaks in the spectra. For Ge(Li) spectra the computational steps are simpler and generally involve (a) spectrum smoothing, (b) peak identification, (c) background subtraction, (d) peak area and multiplet analysis, (e) decay corrections, and (f) calculation of nuclide activities or elemental concentrations. The complexity of the program will depend on the nature of the spectra and the computer available and many computer codes have been proposed (67–71).

The use of digital computers in activation analysis is not restricted to spectrum analysis. Several authors have published details of completely automated analytical procedures (72–76). Such techniques are particularly useful for large numbers of similar samples.

ERRORS AND INTERFERENCES IN ACTIVATION ANALYSIS

Due to the nuclear nature of the method, the NAA technique is subject to few systematic sources of error compared to many chemical or physico-

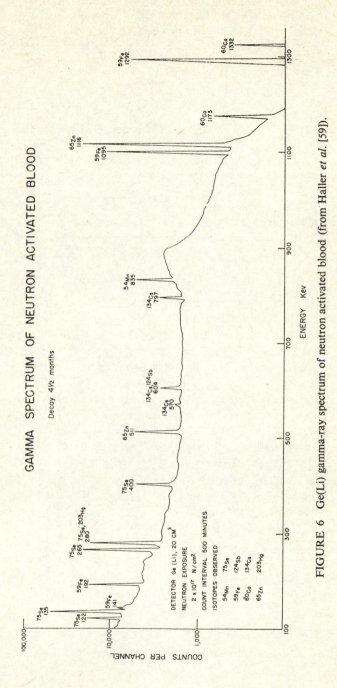

FIGURE 6 Ge(Li) gamma-ray spectrum of neutron activated blood (from Haller *et al.* [59]).

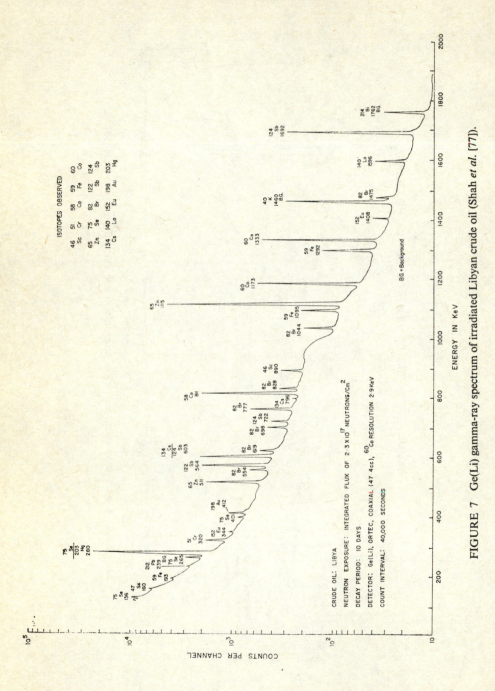

FIGURE 7   Ge(Li) gamma-ray spectrum of irradiated Libyan crude oil (Shah *et al.* [77]).

chemical methods of analysis. There are, however, several sources of error which are associated specifically with NAA and may be divided into errors (or interferences) associated with irradiation and those associated with the measurement of induced radionuclide activities.

Irradiation errors

The errors associated with irradiation involve:

a) neutron flux perturbation in samples or standards and
b) interfering nuclear reactions on other elements which produce the nuclide of interest.

If either the sample or the standard has a high thermal-neutron absorption cross section (σ_a), then the neutron flux in the interior of the sample will be depressed, resulting in a low specific activity of the induced radionuclide in the sample. Fortunately most biological and environmental materials have major elements of low absorption cross sections (e.g. O, N, C, H, Si, Al, etc.) and thermal neutron flux depression is negligible. With environmental materials problems may arise with samples high in chlorine ($\sigma_a = 33.3$ b), e.g. sea water. Neutron flux depression may be observed if concentrated solutions of trace-element standards are employed. Important environmental elements with high absorption cross sections are Cd ($\sigma_a = 2460$ b), Hg ($\sigma_a = 370$ b), Ag ($\sigma_a = 63$ b), In ($\sigma_a = 194$ b), Au ($\sigma_a = 98.8$ b), Co ($\sigma_a = 37.6$ b), and Mn ($\sigma_a = 13.3$ b). Self absorption problems can be eliminated by dilution of the sample or standard such that self absorption is negligible. If this is not practical, as in the case of the determination of trace elements in sea water, then standards must be made possessing the same bulk composition as the samples such that neutron attenuation is the same for both.

During irradiation, the nuclide of interest may be formed by competing nuclear reactions shown by other elements present in the sample. In thermal neutron activation analysis fast neutron reactions may produce the nuclide formed by (n, γ) capture by the element being determined. An example is the determination of cobalt by the $^{59}Co(n, n)^{60}Co$ reaction:

reaction used $^{59}Co(n, \gamma)^{60}Co$

interference $^{60}Ni(n, p)^{60}Co$

interference $^{63}Cu(n, \alpha)^{60}Co$

Thus the presence of Ni or Cu in the sample can interfere in the determination of Co. Such interferences are usually small because most fast-neutron reaction cross sections are much lower than those of (n, γ) reactions and fast-neutron fluxes above the reaction threshold energies may be orders of magnitude lower than the thermal-neutron flux in most reactor irradiation facilities. However, the interference may be significant if the interfering element is a major element in the sample or present at a much higher concentration than the element of interest. In a crude oil containing 344.5 ppm Ni and 0.031 ppm Co, the $^{60}Ni(n.p)^{60}$ contribution to the ^{60}Co activity in a water moderated reactor irradiation position (cadmium ratio = 8) was measured by Shah *et al.* (77) to be 6.7%. Several authors (78, 79) have pointed out the large interference by the $^{35}Cl(n, \alpha)^{32}P$ reaction in the determination of phosphorus in sea water using the $^{31}P(n, \gamma)^{32}P$ reaction. This (n, α) reaction also occurs with thermal neutrons. Such interferences, where significant, may be corrected for by determining the "apparent" concentrations of the elements of interest in irradiated pure samples of interfering elements. The correction thus requires measurement of the concentration of the interfering elements in the samples as well as that of the element of interest.

Other types of interfering reactions are:

a) successive capture interference: $^{63}Cu(n, \gamma)^{64}Cu \xrightarrow{\beta-} {}^{64}Zn(n, \gamma)^{65}Zn$
    reaction used: $^{64}Zn(n, \gamma)^{65}Zn$

b) fission product formation: $^{235}U(n, f)^{140}La$
    reaction used: $^{139}La(n, \gamma)^{140}La$

The successive capture reaction is of negligible significance except where the interfering element is the major element in the sample (e.g. determination of Zn in pure Cu). The ^{235}U fission interference may be significant for nuclides on the maxima of the fission yield-mass curve and for samples high in U but may be corrected for in the same manner as other interfering reactions.

Errors in activity measurements

In NaI(Tl) gamma-ray spectrometry the poor resolution of the detector results in significant overlapping of peaks in complex spectra. With Ge(Li) spectra overlapping peaks are far less common but large errors may result if overlapping-peak energies are very close and unresolvable, e.g. 279.1 keV of ^{203}Hg and 279.3 keV of ^{75}Se. Corrections may be made by obtaining the interfering peak area from that of another peak of the interfering nuclide, provided the relative intensities of the two are known, or have been measured under the experimental conditions employed. For low-energy gamma rays (< 200 keV), all samples and standards must be of similar geometry and similar density to avoid differences in gamma-ray absorption during counting. Care must also be taken to avoid the use of peaks that fall on Compton edges of higher energy gamma rays.

In general, few systematic errors are important in the measurement of radionuclides by Ge(Li) gamma-ray spectrometry.

For samples of high activity corrections must be made for coincidence events and for nuclides of short half-lives, correction for decay during counting must be made.

ACCURACY, PRECISION AND SENSITIVITY OF ACTIVATION ANALYSIS

The accuracy and precision of an analysis will depend on a number of factors, some of which are common to most analytical techniques. Errors in weighing of samples, homogeneity of samples, concentrations of standards, pipetting of standard solutions and computations can all be reduced to insignificant values by careful techniques. For most neutron activation methods the most important factors affecting accuracy are the uniform neutron exposure of samples and standards, neutron self-absorption during irradiation and the measurement of the activities of the induced radionuclides. These sources of error and interferences have been discussed in the previous section. Inaccuracies may also arise due to loss of volatile materials during irradiation (e.g. Hg), adsorption of recoiling ions on the container wall during irradiation or due to transfer of material from the container to the sample but these effects are usually small. For most methods using the comparator technique accuracies of 1–10 per cent may be attained, provided possible sources of error are carefully evaluated.

Precision is also a function of some of the factors affecting accuracy, e.g. weighing errors, sample inhomogeneity, etc., but for trace analysis in environmental materials, the precision of an analysis is often determined by the statistics of activity determination. Depending on the type of materials being analyzed and the concentration of the element measured, relative standard deviations for a single measurement usually fall in the range 1–15 per cent.

Comparison with other analytical methods

In common with most analytical techniques, neutron activation analysis has advantages and disadvantages relative to other methods. The major advantages and disadvantages are listed below.

(*i*) *Advantages*

a) Sensitivity: The method has very high sensitivity for many elements, in some instances to 10^{-10} g (see later).

b) Matrix effects: The fact that nuclear reactions are involved results in the chemical or physical nature of the matrix being unimportant, (exceptions occur where a major matrix element has a high absorption cross section). Thus samples and standards do not have to have similar bulk compositions.

c) Contamination: As the only operations performed on environmental samples prior to irradiation are usually collection and preparation, there is no reagent blank and the possibility of contamination from apparatus or reagents is greatly reduced or eliminated.

d) Multielement technique: For many applications the method is non-destructive and multielement. For example, as many as 28 elements may be determined in air particulates (80). Even where radiochemical separations are necessary, groups of elements can often be separated rather than individual elements.

e) Isotopic ratios: Where an element possesses several stable isotopes, isotopic ratios may be measured in certain cases by activation analysis. Thus $^{64}Zn/^{68}Zn$ ratio has been measured using the $^{65}Zn/^{69m}Zn$ activity ratio relative to standards of known isotopic content (81). Similarly, activation analysis has been used to measure the $^{235}U/^{238}U$ ratio in a number of materials (82).

(ii) *Disadvantages*

a) Not all elements possess suitable radioactive nuclides; either formation cross sections are low or half-lives are very long or very short, resulting in poor sensitivity.

b) In INAA methods for environmental samples, decay periods of up to one month may be necessary to allow the determination of some long-lived nuclides. Hence, the method will have a time lag for some results.

c) No direct information on the chemical state of the element is obtained.

d) Not all laboratories have access to a nuclear reactor.

e) Initial equipment costs for Ge(Li) spectrometry and irradiation facilities are high.

f) Long counting times may limit the number of samples that can be processed.

In many environmental problems, the analyst is interested in high sensitivity methods particularly for the determination of "background" levels of pollutants. In NAA the detection limit for a given element will depend on (a) reaction cross section, σ, (b) reactor neutron flux, (c) half-life of the nuclide produced and irradiation time, (d) decay scheme of the radionuclide, and (e) the method of measurement.

When instrumental Ge(Li) spectrometry is used to measure a given element in a sample the detection limit, for given conditions, will depend on the precision required and to a greater extent on the other elements present in the sample. As the γ-ray peak area for the nuclide of interest is superimposed on a plateau formed by the Compton continuum of other nuclides present, the standard deviation of the area (s_a) will depend on the continuum area on which the peak is superimposed:

$$s_a^2 = s_t^2 + s_b^2$$

where

s_t^2 = total area of peak + continuum

s_b^2 = area of continuum under the peak.

This means that elemental detection limits will depend on the composition of the sample and may vary significantly even among samples of the same type. To indicate analytical sensitivities by INAA, detection limits for a number of elements in human blood (82), crude oil (77, 83), aerosols and air particulates (80) are shown in Table VIII.

If chemical separations are performed, then the detection limits are independent of sample type although the method of measurement will also affect the sensitivity, e.g. Geiger counter, NaI(Tl) or Ge(Li). Table VIII also shows detection limits for elements based on a 1 gm sample, thermal neutron flux of 4.3×10^{12} n.cm^{-2}sec^{-1}, one-hour irradiation, chemical separation of the nuclide of interest and γ-ray measurement using a $3'' \times 3''$ NaI(Tl) detector. These values have been calculated by Yule (84). Meinke (85) has also calculated sensitivities for a neutron flux of 10^{13} n.cm^{-2}sec^{-1} and an irradiation time of 30 days, or saturation.

Table IX shows a comparison of detection limits obtained by activation analysis and detection limits for several other analytical procedures (84–86).

APPLICATIONS

Most applications of activation analysis to environmental problems can be grouped under air or water pollution. The application of neutron activation analysis to the determination of trace elements in biological materials has been comprehensively reviewed recently by Leddicotte (87). An excellent general bibliography of applications of activation analysis has been published by Lutz *et al.* (88) and this bibliography is updated at frequent intervals.

Air pollution

Increasing concern over air quality and the presence of aerosols or air particulates containing toxic elements in air over industrial or urban areas has led to emphasis on very sensitive methods of elemental analysis. Neutron activation analysis has found widespread application to the measurement of aerosols and air particulates because of the high sensitivity of the method, the possibility of measuring many elements in a single sample, and the non-destructive nature of the method. Iddings (89) and Mott (90) have discussed the application of activation analysis and other nuclear techniques to air pollution studies. Instrumental neutron activation analysis using NaI(Tl) spectrometry for analysis of air particulates or aerosols has been proposed by several authors (91–95). Parkinson and Grant (91) measured Al, Na, Cl, I, Mg and Mn in aerosols collected on Whatman #41 filter papers but found it

TABLE VIII

Detection limits (in ppb) for instrumental and radiochemical neutron activation analysis[a]

Element		Instrumental Ge(Li)		Radiochemical	
	Blood (59)[b]	Crude Oil (77, 83)[c]	Air particulates (80)[d]	Yule (84)[e]	Meinke (85)[f]
P	—	—	—	—	1.0
V	—	10	1	0.3	0.05
Cr	10	23	20	450	10
Fe	10	400	1500	10^5	450
Co	1	0.6	2	3.2	1.0
Ni	—	30	1500	180	1.5
Cu	—	100	50	10	0.35
Zn	100	200	100	83	2.0
Ge	—	—	—	29	2.0
As	—	6	40	1.5	0.1
Se	10	20	10	13	2.5
Br	0.5	1	25	4.8	0.15
Mo	—	—	—	47	5.0
Ag	100	—	100	34	5.5
Cd	—	—	—	3.9	2.5
Sb	5	1	80	2	0.2
Au	1	0.1	1	0.07	0.15
Hg	10	4	10	83	6.5
Tl	—	—	—	—	30
Pb	—	—	—	2×10^5	100
U	—	1.5	—	—	0.5

[a] All concentrations based on a 1 gm sample.
[b] Based on irradiation of 2×10^{15} n.cm^{-2}, Ge(Li) spectrometry.
[c] Based on irradiation of 2.3×10^{17} n.cm^{-2}, Ge(Li) detector.
[d] Values for air particulates are ng per 24 hour sampling period based on irradiation time 5 min – 5hr at 2×10^{12} n.cm^{-2} sec^{-1}.
[e] Based on 1hr irradiation at 4.3×10^{12} n.cm^{-2} sec^{-1} – activity measured at end of irradiation with 3″ × 3″ NaI(Tl).
[f] Based on fiux of 10^{13} n.cm^{-2} sec^{-1} and irradiation time of 30 days or saturation. Radionuclide detection limit taken as 40 dps.

necessary to make corrections for the elemental contents of the filter paper. Keane and Fisher (92) used NaI(Tl) spectrometry and a spectrum stripping γ-ray analysis computer code to survey Al, Br, Cl, Mn, Na and V concentrations at weekly intervals in selected areas of Britain. NaI(Tl) spectrometry has also been extensively used by Brar and co-workers (93, 94) to analyze air

TABLE IX

Comparison of sensitivity (in ng/g) of several analytical methods of trace element determination

Element	NAA (66)	Photometric[a]	Atomic absorption (86)[b]	Emission spectrography (85)
P	1.0	1.0	—	2×10^4
V	0.05	10	20	50
Cr	10	7.0	5.0	50
Fe	450	200	10	500
Co	1.0	3.0	7.0	500
Ni	1.5	4.0	10	100
Cu	0.35	2.0	5.0	200
Zn	2.0	100	2.0	2000
Ge	2.0	20	1000	—
As	0.1	10	500	5000
Se	2.5	200	500	—
Br	0.15	—	—	—
Mo	5.0	10	100	50
Ag	5.5	5.0	5.0	100
Cd	2.5	3.0	1.0	2000
Sb	0.2	4.0	200	5000
Au	0.15	5.0	10	200
Hg	6.5	5.0	1.0[c]	5000
Tl	30	20	200	200
Pb	100	6.0	10	50
U	0.5	300	1.2×10^4	1000

[a] Based on 1 ml cell, 1 cm path length.
[b] Based on 1 ml solution needed for analysis.
[c] Flameless atomic absorption value.

particulates in surface air from the Chicago metropolitan area. These authors used short reactor irradiations to measure Al, Br, Cl, Mn, Na and V and long irradiations to measure Hg, Cr, Zn, Fe, Sb, Sc, and Co. The complex γ-ray spectra collected were resolved by a simultaneous equations code. Tuttle *et al.* (95) used NaI(Tl) spectrometry and least-squares spectrum resolution to determine Al, Br, Ca, Cl, Mn, Na and V in aerosols and particulates from a heavily industrialized area (Cincinnati) and from an area of little industry (Columbia, Mo.). The introduction of Ge(Li) detectors has resulted in increased use of INAA in air pollution studies since a larger number of elements can be measured than with NaI(Tl) spectrometry. Zoller and Gordon (96) describe analytical methods, grouped by half-life of the induced radionuclide, for 24 elements in aerosols collected from 20–50 m^3 of air. These authors also used a low energy photon spectrometer to measure rare earths, [75] Se and [181] Hf in addition to Ge(Li) γ-ray spectrometry. In a later paper Gordon et al. (97) discuss the advantages of INAA over emission spectrography and nondispersive X-ray fluorescence. They demonstrated the possibility of using INAA to "fingerprint" pollution sources on the basis of trace element contents of the aerosols and source materials and noted the correlation of V in aerosols with the burning of residual fuel oils and Pb and Br with auto-exhaust products from lead-containing gasolines. Dams et al. (80, 98) investigated different techniques of sample collection and proposed use of high-vacuum pumps with low trace-element polystyrene filters for collection of aerosols and air particulates. A sequential INAA scheme based on half-lives of neutron induced activities was described for measuring Al, V, Cu, Mg, Ca, Ti, S, Cl, Br, Mn, In, (short half-life group), and Zn, Br, As, Ga, La, Sm, Eu, Sb, W, Au, Sc, Cr, Cd, Fe, Ni, Se, Ag, Ce, Hg, Th, (long half-life group). Ge(Li) spectra were collected on 7-track magnetic tape and all data reduction was performed by computer codes. In many samples up to 33 elements could be determined. Dams et al. (99) point out that the high sensitivity of INAA for certain elements allows sampling of 4 m^3 of air at 90 minute intervals for 15–20 elements which permits measurement of diurnal trace-element variations. In a study of trace elements in air particulates over industrialized N.W. Indiana, these authors showed that Se and S correlated with fossil fuel combustion; Fe, Mn, Zn, Sb, Cr, W, Co, Sc, La, Ce, Th, Ca and Mg were linked to steel-making activities whereas Na, K, Sm, and Eu were not unique to industrialized areas. Dams et al. (80) also made a study of correlations of trace elements with particle-size fractions collected on 7-stage impactor sampler to study chemical processes occurring in the environment. Pillay and Thomas (100) used INAA, Ge(Li) spectrometry and X-ray spectrometry to measure up to 19 elements in 200 samples collected from the Buffalo, N.Y. area. These authors used sequential sampling techniques and liquid traps for collection of fine aerosols ($< 0.1 \mu$) not trapped by filters and found that approximately 50 per cent of the Se passed the filters and was retained in the liquid traps. These authors also used high-flux pulsing to measure Pb by the ^{206}Pb $(n, \gamma)^{207m}$Pb (0.8 sec) reaction.

Several applications of activation analysis to determine air particulate or aerosol sources have been reported. Rancitelli and Perkins (101) used INAA to measure Ag, Co, Cr, Fe, Sb, Se and Zn in the troposphere and lower stratosphere. Air particulates were collected between 5 and 15 km altitudes and the atmospheric trace-element levels were correlated with these in metropolitan aerosols, sea water, the earth's crust and unpolluted rain water. The authors found that Fe, Sc, and Co were consistent with a crustal origin but that Ag, Cr, Zn and Sb were too high to be of natural origin. Dudey *et al.* (102) using radiochemical group separations followed by Ge(Li) spectrometry determined 20 elements in marine aerosols. They found that Zn, Hg, Cd, and Se were partly from pollution sources whereas the rare earths were of terrestrial origin. Rahn *et al.* (103) used INAA to measure 15 elements in aerosol samples collected at 2-hour intervals for a 58 hour period in Livermore, California. They found a cyclic behavior for trace elements with Br correlating with automobile exhaust, Na and Cl due to aged marine air and Mn, Al, Fe, Sc being due to continental weathering. Deviations from these patterns were shown by Zn, Sb and V and local pollution sources were indicated. Several studies have been made of halogens and Pb in terrestrial and marine air using β-counting (104, 105) or NaI(Tl) spectrometry (106–108).

To determine the sources of industrial air pollution it is important that the concentrations of pollutant elements in possible source materials be known. Shah *et al.* (77, 79) have measured 23 elements in crude oils by INAA and Ge (Li) spectrometry. Filby and Shah (109) have shown that trace metals such as Ni, V, Hg, As, Cu, Co, Se and Zn are concentrated in the high molecular weight fractions of crude oils and these elements will also accumulate in residual fuel oils. Considerable variation was noted in trace element abundances in oils of different geographical locations and "fingerprinting" of oil-burning sources appears possible. Several other applications of NAA to petroleum fuels have been reported (110–112). Ruch (113) has used neutron activation followed by radiochemical separation to determine Hg in Illinois coals and found an average of 0.18 ppm Hg in 55 coals. Pillay *et al.* (114) used NAA and radiochemical separation of Se to determine Se levels in atmospheric pollutants and in fossil fuels. They found that measurement of the Se/S ratio could be used to determine whether an air pollution source was combustion of crude oil or coal. Also, the Se level was found to be a useful indicator of total sulfur pollutants in the atmosphere.

The aquatic environment

Activation analysis plays an important part in the measurement of polutants in the aquatic environment. In order to understand the nature and amounts of pollutants added to the hydrosphere and their ultimate fate, it is important that the geochemical cycle of each element in both marine and freshwater environments be known. The geochemical cycles of some elements have been established although many details need to be elucidated. For many elements the geochemical cycles are poorly defined, often because of unsatisfactory analytical data and the difficulty of measuring the very low concentrations of many elements in the hydrosphere and aquatic organisms. Very little is presently known of the distribution and ultimate fate of most freshwater food chains. It is in this area of study that activation analysis finds widespread use.

Activation analysis has been used extensively in both marine and freshwater investigations and for many problems requiring high sensitivity it is the only practical method of analysis. Hogdahl (115) has briefly reviewed the literature on applications of activation analysis to oceanography and Lutz (116) has compiled a bibliography of papers published up to 1970. Kruger (117) has discussed the application of various nuclear techniques to hydrology and reviews the use of NAA as a technique for measuring pollutants and for measuring stable isotopes added to aquatic systems as tracers for water or material migration.

The very low contents of many trace elements in seawater (many are present at less than 0.1 ppm) and the high Na, Cl, and Br contents of seawater restricts the measurement of induced radionuclides to those of long half-life, unless chemical separations of the nuclides of interest from the very high ^{24}Na, ^{38}Cl and ^{82}Br activities can be made. However, to obtain the required sensitivity for trace elements large neutron exposures (ϕt) are necessary and very high ^{24}Na, ^{38}Cl, ^{32}P, and ^{82}Br radiation dose rates also create handling problems—hence, a long decay period is usually necessary. One solution to this problem is to concentrate the trace elements from a large volume of sea water into a smaller volume with low Na, Cl, and Br contents. Fukai (118) has discussed pre-irradiation procedures and points out the possibilities of contamination through apparatus and reagents and loss of trace elements by adsorption on surfaces. Acidification of samples immediately was recommended in those cases where only the total elemental content is to be measured. Several authors have used pre-concentration prior to irradiation by ion exchange (31, 119) or evaporation (32).

Few systematic studies of the relative merits of NAA and other analytical techniques used in chemical oceanography have been made. Boudin *et al.* (120) have, however, compared NAA, atomic absorption spectrophotometry, and isotope dilution analysis for several elements. They found good agreement among methods for Lu, Ni and Mn, and NAA was superior to submicrogram amounts of other elements considered. To reduce contamination and losses in all techniques, they proposed freezing of seawater samples immediately after collection.

Several multielement NAA methods have been proposed for seawater. Schutz and Turekian (121) determined 18 elements in seawater collected from different, worldwide localities to investigate geographical and vertical concentration variations. In their methods, the water samples were freeze-dried and the resultant salts irradiated. After irradiation for 150 hours at $2-6 \times 10^{12}$ n.cm^{-2}sec^{-1} and fusion of the sample with Na_2CO_3 plus Na_2O_2 each element was separated by precipitation and ion-exchange techniques. NaI(Tl) spectroscopy was used to measure the separated radionuclides. The authors obtained satisfactory analyses for Rb, Cs, Ba, Au, Sb, Se, Ag, Co and Ni and adsorption losses were investigated for Zn, Cr, Sb, Co, Ag, Cs, and Se. Piper and Goles (32) used INAA and Ge(Li) spectrometry to measure Co, Cr, Fe, Rb, Sr, Cs, Sc, Sb, and Zn in freeze-dried sewater samples. Robertson *et al*. (122, 123) have developed highly sophisticated INAA procedures for the determination of Sr, Cs, U, Co, Sb, Zn, Fe, Sc, Hg and Ag in seawater, marine organisms and sediments. They used high-neutron-flux irradiation of samples followed by Ge(Li) spectroscopy or multidimensional NaI(Tl) spectrometry. Robertson and Prospero (123) investigated trace element contents in a large number of Atlantic waters and related the highly variable concentrations of Co, Zn and Fe to biological uptake. Radiochemical methods have been published for In (119), Tl (31), As (124), Sb (125), Cr (126), and Mo (126) in seawater. Marine organisms as potential concentrators of pollutants have been analyzed by NAA for several elements. Rancitelli *et al*. (117, 128) used INAA to measure 14 elements in marine organisms and Pacific salmon tissues and related the values obtained to seawater concentrations. Fukai and Meinke (129) used radiochemical separations after neutron irradiation to measure V, As, Mo, W, Re and Au and Hamaguchi *et al*. (130) measured Hg, Cu, and As in several marine organisms. The use of ^{252}Cf sources in marine investigations has been reported by Wiggins *et al*. (131) and John (12). John (12) used reverse osmosis to concentrate trace elements in seawater and the concentrate was irradiated in a ^{252}Cf neutron source. Detection limits for 12 elements using a 5 mg ^{252}Cf source ranged from 0.001 mg/l (Ag, In, Sc, V) to 0.08 mg/l (Hg). These detection limits are for interference-free cases and represent "ideal" situations. At present the technique does not appear to be competitive in terms of sensitivity with reactor activation analysis for trace elements in seawater.

More is perhaps known concerning pollutants in the freshwater environment because pollution is usually of more direct concern (no large dilution factor) and more obvious. The use of activation analysis has grown rapidly in this area, particularly for the investigation of mercury pollution (see next section). Merlini and associates (132–135) have used NAA to measure Zn, Mn, Fe, Co, Cu in some Italian freshwater ecosystems to determine the distribution of these elements in the food chains and the use of NAA in such studies is certain to grow. The analysis of freshwaters from rivers, lakes and springs has received considerable attention. Leddicotte and Moeller (136) and Blanchard *et al*. (137) analyzed drinking water from 7 U.S. localities in the first comprehensive application of NAA to water analysis. These authors measured 18 elements by NaI(Tl) spectroscopy after radiochemical separations following irradiation. Funk *et al*. (138) used Ge(Li) spectrometry to determine Na, Cr, Sc, Co, Ba, Sb, Au, Fe, Th, Zn, Cs, Rb, Hg, U and Br following the irradiation of 30 ml samples. Removal of ^{24}Na on HAP (57) allowed the simultaneous determination of Cu, Ga, La, and K which have nuclides with half-lives similar to ^{24}Na. Both lakewaters and algae specimens were analyzed and samples were irradiated and counted in the collection container (polyethylene) to eliminate adsorption problems. Container blanks were run for all elements. Rancitelli and Tanner (139) and Perkins and Rancitelli (29) have measured 19 elements in river water by a comprehensive analytical scheme. Water samples were collected, frozen and irradiated frozen to prevent contamination and adsorption on surfaces. Following irradiation the sample was melted, filtered and element groups separated by adsorption on Dowex-1 in OH^-, S^{--} and CO_3^{--} forms. In addition to Ge(Li) spectrometry, a Ge(Li)-NaI(Tl) coincidence/anticoincidence spectrometer was used to enhance spectral peak-to-Compton ratios. Schmidt (140) demonstrated the use of radiochemical NAA as a survey technique in water quality evaluation in which monthly samples were taken at different localities in Germany. Schmidt (140) calculated the detection limits to be 10^{-5} μg/l for Br, Ag, In, Sb and Cs and 10^{-4} μg/l for Zn, Se, Rb, Ba and Ce. Kharkar et $al.$ (141) used NAA to determine the global stream supply of Ag, Mo. So, Cr, Ce, Co, Rb, and Cs to the oceans. Several other multielement NAA techniques of freshwater analysis have been reported (142–146). Wahlgren et $al.$ (147) have pointed out the utility of combined analytical techniques in water quality evaluation. They used spark source mass spectrometry in conjunction with the semi-automated NAA technique of Edgington and Lucas (74). The use of two techniques provides valuable cross checking of data in addition to the methods supplementing each other. Ljunggren et $al.$ (148) used NAA to measure the contents of the heavy metal pollutants As, Hg, Cd in Swedish lake, river, and drinking waters in addition to aquatic organisms. Sensitivities claimed were 0.01 ng/g for Hg, 0.5 ng/g for As, and 1 ng/g for Cd. Other applications of NAA of more restricted scope include the measurement of phosphate (78), Cr (149), Au (150, 151), V (152), and Mg, Sr and Ni (153).

Mercury pollution

Many of the studies quoted in previous sections have been concerned with multielement analysis and have included Hg among the elements determined. Mercury pollution has aroused public concern and the serious problem arises principally from the disposal of industrial Hg wastes and the use of Hg fungicides in agriculture. Mercury readily forms stable methyl mercury of the type (CH_3HgX; X = anion) derivatives in the aquatic environment and these enter food chains to a remarkable degree. Methyl mercury ion is extremely toxic to vertebrates and affects the nerve cells of the brain. Since

1955 many studies of the environmental behaviour and hazards of Hg have been made and form the subject of recent reviews (154–156). The analytical chemistry of Hg is complicated by the volatility of the metal, the ease of disproportionation of Hg(I), and the presence of different chemical states, e.g. inorganic and organo mercury derivatives, in many biological and environmental materials. Unless extreme care is taken, Hg may be volatilized and lost during such common analytical procedures as dissolution or distillation of solutions containing inorganic Hg. The stability and volatility of dimethyl and other dialkyl or diaryl Hg compounds can result in even more serious losses. Mercury is readily lost from very dilute solutions of Hg(II) (157, 158) and reduction of Hg(II) and disproportionation of Hg(I) has been proposed to explain this behaviour (157). Also oven drying, and low temperature ashing lead to Hg losses from biological material of up to 98% (34, 40). Sediments in which Hg is present predominately as inorganic Hg(II) lose insignificant amounts under these conditions. In neutron activation methods, significant losses of Hg from standard solutions encapsulated in polyethylene vials have been reported (38, 40, 158) but no losses occur from quartz vials. If samples and standards are irradiated in quartz vials to prevent losses during irradiation, and if decomposition of samples is carried out in a closed system for radiochemical activation analysis, then accurate Hg analyses may be made by NAA.

Filby *et al.* (159) have described an INAA method for Hg in biological and environmental materials using measurement of ^{203}Hg by Ge(Li) spectrometry. For human blood the detection limit was 0.0035 μg/g and detection limits for this instrumental method for most materials were determined by the ^{75}Se content of the irradiated sample (overlap of 279 keV gamma rays from ^{75}Se and ^{203}Hg. Several authors have published radiochemical NAA methods for Hg in which losses of Hg during chemical processes are reduced to a minimum or eliminated. Sjostrand (160) used a radiochemical technique to separate ^{197}Hg after irradiation. The sample was wet-ashed with HNO_3-H_2SO_4 in a closed system and Hg electrodeposited on Au-foil cathodes prior to counting of the 70 keV (X-ray–γ-ray) group of ^{197}Hg. No losses were observed in this procedure. Kosta and Byrne (161) used high temperature ignition (400°C) of the irradiated sample, removal of volatile elements on Ag-coated quartz wool at 300°C followed by a retention of Hg metal on Se impregnated paper as HgSe. Rottschafer *et al.* (162) and Jones *et al.* (163) used anion exchange retention of $HgCl_4^{--}$ after oxidative dissolution of biological samples followed by direct counting of ^{197}Hg on the resin. These authors report a sensitivity of 3 ppb Hg in fish tissues. Bate (40) used quartz vials for all irradiations of biological materials and decomposition of the sample was carried out with an HNO_3-H_2SO_4-NH_4VO_3 mixture under reflux conditions to trap all Hg species. After dissolution of the sample

HgS was precipitated with S^{--} ion and counted for ^{197}Hg. Comparisons were made with NAA and atomic absorption and the latter method was found to give low results, indicating probable losses of Hg during sample processing. Pillay *et al.* (34) have carefully investigated the analytical manipulations of biological materials and conclude that best results were obtained by irradiation of wet tissue together with aqueous standards in quartz vials. After irradiation H_2SO_4-$HClO_4$ ashing was used for sample dissolution and HgS precipitated by H_2S. After purification of the Hg, the metal was electrodeposited on Au foils and the ^{197}Hg counted. These authors present data on comparisons of NAA, flameless atomic absorption, flame atomic absorption and spectrophotometric methods for Hg determination. Considerable variations were obtained among methods for identical samples and the authors concluded that losses occurred in techniques other than NAA (unless specific precautions were taken) and that NAA was superior to other methods.

The problem of the Hg content of foods for human consumption has received much attention, particularly in Scandinavia and Canada. Christell *et al.* (158) used INAA to measure Hg in waters, birds, eggs, and fish in a large scale screening programme. The methods used were similar to those developed by Sjostrand (160). Jervis and Tiefenbach (38) and Rayudu *et al.* (164) used both INAA with Ge(Li) spectrometry to measure ^{203}Hg for high-level samples and radiochemical NAA and NaI(Tl) spectrometry for low-level samples. These authors observed no losses of Hg during irradiation of samples in polyethylene, except for fish samples containing dimethyl Hg. A variety of Canadian foods were analyzed and an average daily intake of Hg was calculated from the data. The authors also proposed the analysis of human hair as a technique for monitoring Hg exposures. Guinn and Kishore (165) have analyzed tuna samples by INAA using Ge(Li) determination of ^{197}Hg and found concentrations of Hg of between 0.262–0.461 μg/g (wet weight). Levels in tuna and swordfish higher than the U.S. Food and Drug Administration "tolerance" limit of 0.5 ppm Hg have led to concern that Hg pollution of the ocean is becoming significant. Miller *et al.* (166) have shown, however, that 7 tuna samples collected between 1878 and 1909 showed an average Hg content of 0.95 μg/g (dry weight) compared to recent tuna values of 0.21 μg/g and the authors conclude that there is no evidence for increased Hg levels in marine fish due to industrial Hg pollution.

Activation analysis for Hg has also been used in toxicological problems (167, 168), in the study of recent sediments of L. Michigan (169) and in the biological effects of a Hg mine on regional ecology (170).

References

1. P. Kruger, *Principles of Activation Analysis* (John Wiley, New York, 1971).
2. D. DeSoete, R. Gijbels, and J. Hoste, *Neutron Activation Analysis* (John Wiley, New York, 1971).

3. J. Hoste, J. Op DeBeeck, R. Gijbels, F. Adams, P. Van der Winkel, and D. DeSoete, *Activation Analysis* (Chemical Rubber Co. Press, Cleveland, Ohio, 1971).
4. M. Rakovic, *Activation Analysis* (Iliffe Books Ltd., London, 1970).
5. R. H. Filby, A. I. Davis, and K. R. Shah, *Radiochem. Radioanal Letters* **5**, 9 (1970).
6. C. M. Lederer, J. M. Hollander, and I. Perlman, *Table of Isotopes* (John Wiley, New York, 1967) 6th ed.
7. R. H. Filby, A. I. Davis, K. R. Shah, G. G. Wainscott, W. A. Haller, and W. A. Cassatt, *Gamma Ray Energy Tables for Neutron Activation Analysis* (Washington State Univ., Pullman, 1970).
8. B. T. Kenna and F. J. Conrad, *SC-RR*-66-229 (*TID*-4500; *UC*-34) *Sandia Corporation Report* (1966).
9. R. S. Rochlin, *Nucleonics* **17**, 54 (1959).
10. R. W. Perkins, L. A. Rancitelli, J. A. Cooper and R. E. Brown, *Nucl. Appl. Tecnol.* **9**, 861 (1970).
11. J. L. Crandall, *Isotopes Radiat. Technol.* **7**, 306 (1970).
12. J. John, Proc. American Nuclear Society Topical Meeting, *Nuclear Methods in Environmental Research*, Columbia, Mo., Aug. 23–24, 1971, 16 (Univ. Missouri, 1971)
13. F. E. Senftle, *Marine Technol. Soc. J.* **3**, 9 (1969).
14. P. E. Quittner and R. E. Wainerdi, *At. Energy Rev.* **8**, 361 (1970).
15. J. A. Cooper, *USAEC Report BNWL-SA*-3603 (1971).
16. L. Salmon and M. G. Creevy, *Nuclear Techniques in Environmental Pollution* Proc. Symposium, Salzburg, 26–30 Oct., 1970, 47 (International Atomic Energy Agency, Vienna, 1971).
17. C. J. Thompson, *Nucl. Appl.* **6**, 559 (1969).
18. J. A. Cooper and R. W. Perkins, *USAEC Report BNWL-SA*-3972 (1971).
19. J. A. Cooper and R. W. Perkins, *USAEC Report BNWL-SA*-3527 (1971).
20. J. A. Cooper, L. A. Rancitelli, and R. W. Perkins, *J. Radioanal. Chem.* **6**, 147 (1970).
21. J. A. Cooper, *USAEC Report BNWL-SA*-3575 (1970).
22. R. L. Currie, R. McPherson, and G. H. Morrison, *N.B.S. Spec. Publ.* 312 **2**, 1062 (1969).
23. P. L. Phelps, *USAEC Report UCRL*-70544 (1967).
24. B. A. Euler, D. F. Covell, and S. Yamamoto, N.B.S. Spec. Publ. 312, **2**, 1081 (1969).
25. N. A. Wogman, R. W. Perkins, and J. H. Kaye, *Nucl. Instr. Methods* **74**, 197 (1969).
26. D. E. Robertson, *Anal. Chem.* **40**, 1067 (1968).
27. D. E. Robertson, *Anal. Chim. Acta* **42**, 533 (1968).
28. J. L. Nelson, R. W. Perkins, J. M. Nielsen, and W. L. Haushild, *Disposal of Radioactive Wastes into Seas, Oceans and Surface Waters*, Proc. Symposium, Vienna 16–20 May 1966, 161 (International Atomic Energy Agency, Vienna, 1966).
29. R. W. Perkins and L. H. Rancitelli, Proc. American Nuclear Society Topical Meeting, *Nuclear Methods in Environmental Research*, Columbia, Mo., Aug. 23–24, 1971, 47 (Univ. of Missouri, 1971).
30. R. H. Filby, Unpublished data.
31. A. D. Matthews and J. P. Riley, *Anal. Chim. Acta* **48**, 25 (1969).
32. D. Z. Piper and G. G. Goles, *Anal. Chim. Acta* **47**, 560 (1969).
33. D. Brune, *Radiochim. Acta.* **5**, 14 (1966).
34. K. K. S. Pillay, C. C. Thomas, Jr., J. A. Sondel, and C. M. Hyche, *Anal. Chem.* **43**, 1419 (1971).
35. R. H. Filby, J. O. Schmidt, Unpublished data.
36. R. H. Filby, *Anal. Chim. Acta* **31**, 434 (1964).

37. D. Gibbons, Communication to WHO/FAO/IAEA International Discussion, *Mercury Contamination of the Biosphere* (International Atomic Energy Agency, Vienna, 1967).

38. R. E. Jervis and B. Tiefenbach, Proc. American Nuclear Society Topical Meeting, *Nuclear Methods in Environmental Research*, Columbia, Mo., Aug. 23–24, 1971, 188 (Univ. Missouri, 1971).

39. L. C. Bate, *Radiochem. Radioanal. Letter* **6**, 139 (1971).

40. L. C. Bate, Proc. American Nuclear Society Topical Meeting *Nuclear Techniques in Environmental Research*, Columbia, Mo., Aug 23–24, 1971, 197 (Univ. Missouri, 1971).

41. F. J. Flanagan, *Geochim. Cosmochim. Acta.* **33,** 81 (1969).

42. H. J. M. Bowen, *Analyst* **92,** 124 (1967).

43. H. J. M. Bowen, Standard Materials: in *Advances in Activation Analysis*, J. M. A. Lenihan, and J. Thomson, eds. (Academic Press, New York, 1969), vol. 1, 101–114.

44. H. L. Finston and J. Miskeel, *Ann. Rev. Nucl. Sci.* **5**, 269 (1955).

45. H. Frieser and G. H. Morrison, *Ann Rev. Nucl. Sci.* **9**, 221 (1959).

46. I. H. Qureshi, L. T. McClendon, and P. D. LaFleur, *N. B. S. Spec. Publ.* 312, vol. 1, 666 (1969).

47. G. C. Goode, C. W. Baker, and N. M. Brooke, *Analyst* **94,** 728 (1969).

48. A. Alian and A. Haggag, *Talanta* **14,** 1109 (1967).

49. G. Albouin, J. Diebolt, E. Junod, and J. Laverlochere, *Proc.* 1965 *Int. Conf. Modern Trends in Activation Analysis* April 19–22, 1965, College Station, Tex., 344 (Texas A & M, College Station, 1965).

50. K. Samsahl, P. Wester, and O. Landstrom, *Anal. Chem.* **40,** 181 (1968).

51. J. R. DeVoe, *Report NAS-NS* 3108 National Academy of Sciences, Wash. D. C. (1962).

52. K. Samsahl, *Anal. Chem.* **39,** 1480 (1967).

53. F. J. Berlandi, Neutron *Activation Electrodeposition Techniques* (Thesis, Univ. of Michigan, Ann Arbor, Mich., 1966).

54. J. Op de Beeck and J. Hoste, *Acta Chim. Acad. Sci. Hung.* **53,** 137 (1967).

55. E. Cerrai and G. Ghersini, *J. Chromatogr.* **24,** 383 (1966).

56. W. Bock-Werthmann and W. Schulz, *Proc.* 1965 *Int. Conf. Modern Trends in Activation Analysis*, College Station, Tex., April 19–22, 120 (Texas A & M College, College Sta., 1965).

57. F. Girardi and E. Sabbioni, *J. Radioanal. Chem.* **1,** 168 (1968).

58. F. Girardi, R. Pietra, and E. Sabbioni, *EURATOM Report EUR*-4287e (1969).

59. W. A. Haller, R. H. Filby, and L. A. Rancitelli, *Nucl. Appl.* **6,** 365 (1969).

60. D. F. Covell, *Anal. Chem.* **32,** 1086 (1960).

61. H. P. Yule, *N. B. S. Spec. Publ.* 312 vol. 2 1115 (1969).

62. S. Sterlinski, *Anal. Chem.* **42,** 151 (1970).

63. H. P. Yule, *Anal. Chem.* **40,** 1480 (1968).

64. P. A. Baedecker, *Anal. Chem.* **43,** 405 (1971).

65. P. Quittner, *Anal. Chem.* **41,** 1504 (1969).

66. E. Schonfeld, A. H. Kibbey, and W. Davis, Jr., *Nucl. Instrum. Meth.* **45,** 1 (1966).

67. M. A. Mariscotti, *Nucl. Instrum. Method* **50,** 309 (1967).

68. R. Gunnink, H. B. Levy, and J. B. Niday, *USAEC Report UCID*-15140 (1967).

69. T. Harper, T. Inouye, and N. C. Rasmussen, *USAEC Report MITNE*-97 (1968).

70. J. N. Hamawi and N. C. Rasmussen, *USAEC Report MITNE*-107 (1968).

71. V. Barnes, *I.E.E.E. Trans. Nucl. Sci.* NS-15 437 (1968).

72. F. Girardi, G. Guzzi, G. Dicola, W. Becker, and A. Termanini, *N.B.S. Spec. Publ.* 312 vol. 2, 1111 (1969).
73. R. E. Wainerdi, L. E. Fite, D. Gibbons, W. W. Wilkins, P. Jimenez, and D. Drew, *Radiochemical Methods of Analysis*, Proc. Conf. Salzburg, 19–23 Oct. 1964, Vol. 2 164 (International Atomic Energy, Vienna, 1965).
74. D. N. Edgington and H. F. Lucas, *J. Radioanal. Chem.* **5**, 233 (1970).
75. T. B. Pierce, R. K. Webster, R. Hallett, and D. Mapper, *N.B.S. Spec. Publ.* 312 vol. 2 1116 (1969).
76. J. I. Trombka and R. L. Schmadebach, *N.B.S. Spec. Publ.* 312 vol. 2 1097 (1969).
77. K. R. Shah, R. H. Filby, and W. A. Haller, *J. Radioanal. Chem.* **6**, 413 (1970).
78. H. E. Allen and R. B. Hahn, *Environ. Sci. Tech.* **3**, 844 (1969).
79. W. T. Mullins, J. F. Emery, L. C. Bate, and G. W. Leddicotte, *USAEC Report ORNL*-3397 105 (1963).
80. R. Dams, K. Rahn, G. Nifong, J. Robbins, and J. Winchester, Proc. American Nuclear Society Topical Meeting *Nuclear Methods in Environmental Research*, Columbia, Mo., Aug. 23–24, 1971 (Univ. Missouri, 1971).
81. R. H. Filby, *Anal. Chem.* **36**, 1597 (1964).
82. M. Mantel, J. Gilat, and S. Amiel, *N.B.S. Spec. Publ.* 312 vol. 2 79 (1969).
83. K. R. Shah, R. H. Filby, and W. A. Haller, *J. Radioanal. Chem.* **6**, 185 (1970).
84. H. P. Yule, *Anal. Chem.* **37**, 129 (1965).
85. W. W. Meinke, *Science* **121**, (1955).
86. L. L. Lewis, *Anal. Chem.* **40**, (12) 28A (1968).
87. G. W. Leddicotte, *Methods Biochem Anal.* **19**, 345 (1971).
88. G. J. Lutz, R. J. Boreni, R. S. Maddock, and W. W. Meinke, *N.B.S. Technical Note* 467, parts 1 and 2 (National Bureau of Standards, 1971).
89. F. A. Iddings, *Environ. Sci. Technol.* **3**, 132 (1969).
90. W. E. Mott, *Nuclear Techniques in Environmental Pollution*: Proc. Symposium, Salzburg, Oct. 26–30, 1970, 3 (International Atomic Energy Agency, Vienna, 1971).
91. T. F. Parkinson and L. G. Grant, *Nature* **197**, 479 (1963).
92. J. R. Keane and E. M. R. Fisher, *Atmos. Environ.* **2**, 603 (1968).
93. S. S. Brar, D. M. Nelson, E. L. Kanabrocki, C. E. Moore, C. D. Burnham, and D. M. Hattori, *Environ. Sci. Technol.* **4**, 50 (1970).
94. S. S. Brar, D. M. Nelson, J. R. Kline, P. F. Gustafson, E. L. Kanabrocki, C. E. Moore, and D. M. Hattori, *J. Geophys. Res.* **75**, 2939 (1970).
95. R. F. Tuttle, J. R. Vogt, and T. F. Parkinson, *Nuclear Techniques in Environmental Pollution*: Proc. Symposium, Salzburg, Oct. 26–30, 1970, 119 (International Atomic Energy Agency, Vienna, 1971).
96. W. H. Zoller and G. E. Gordon, *Anal. Chem.* **42**, 257 (1970).
97. G. E. Gordon, W. H. Zoller, E. S. Gladney, and A. G. Jones, Proc. American Nuclear Society Topical Meeting *Nuclear Methods in Environmental Research*, Columbia, Mo., Aug. 23–24, 1971, 30 (Univ. Missouri, 1971).
98. R. Dams, J. A. Robbins, K. A. Rahn, and J. W. Winchester, *Anal. Chem.* **42**, 861 (1970).
99. R. Dams, J. A. Robbins, K. A. Rahn, and J. W. Winchester, *Nuclear Techniques in Environmental Pollution*: Proc. Symposium, Salzburg, Oct. 26–30, 1970, 139 (International Atomic Energy Agency, Vienna, 1971).
100. K. K. S. Pillay and C. C. Thomas, Jr., *J. Radioanal. Chem.* **7**, 107 (1971).
101. L. A. Rancitelli and R. W. Perkins, *J. Geophys. Res.* **75**, 3055 (1970).
102. N. D. Dudey, L. E. Ross, and V. E. Noshkin, *N.B.S. Spec. Publ.* 312 vol. 1, 55 (1969).

103. K. Rahn, J. J. Wesolowski, W. John, and H. R. Ralston, *J. Air. Poll. Contr. Assoc.* **21**, 406 (1971).

104. R. A. Duce and J. W. Winchester, *Radiochim. Acta.* **4**, 100 (1065).

105. R. M. Loucks, J. W. Winchester, W. R. Matson, and M. A. Tiffany, *N.B.S. Spec. Publ.* 312 vol. 1 36 (1969).

106. R. L. Lininger, R. A. Duce, J. W. Winchester, and W. R. Matson, *J. Geophys. Res.* **71**, 2457 (1966).

107. J. Winchester, W. H. Zoller, R. A. Duce, and C. S. Benson, *Atmosph. Environ.* **1**, 105 (1967).

108. P. E. Wilkness and R. E. Larsen, *Nuclear Techniques in Environmental Pollution*: Proc. Symposium, Salzburg, 26–30 Oct. 1970, 159 (International Atomic Energy Agency, Vienna, 1971).

109. R. H. Filby and K. R. Shah, Proc. American Nuclear Society Topical Meeting *Nuclear Methods in Environmental Research*, Columbia, Mo., Aug. 23–24, 1971, 86 (Univ. Missouri, 1971).

110. D. J. Veal, *Anal. Chem.* **38**, 1080 (1966).

111. H. A. Braier and W. E. Mott, *Nucl. Appl.* **2**, 44 (1966).

112. G. Palmai, L. Vajta, I. Szabenyi, and G. Toth, *Period. Polytech. Chem. Eng.* **13**, 99 (1969).

113. R. R. Ruch, H. J. Gluskoter, and E. J. Kennedy, *Env. Geol. Notes No.* 43 (1971).

114. K. K. S. Pillay, C. C. Thomas, Jr, and J. A. Sondel, *Environ. Sci. Technol.* **5**, 74 (1971).

115. O. Högdahl, in *Activation Analysis in Geochemistry and Cosmochemistry* (Universitetsforlaget, Oslo, 1971) A. O. Brunfelt and E. Steinnes, eds. 301.

116. G. J. Lutz, *N.B.S. Tech. Note* 534 (1970).

117. P. Kruger, Proc. American Nuclear Society Topical Meeting *Nuclear Methods in Environmental Research*, Columbia, Mo., Aug. 23–24, 1971, 118 (Univ. Missouri, 1971).

118. R. Fukai, Proc. International Symposium *Applications of Neutron Activation Analysis in Oceanography*, Brussels, 1968, 41 (Inst. Roy. Sci. Nat. de Belgique, 1968).

119. A. D. Matthews and J. P. Riley, *Anal. Chim. Acta* **51**, 287 (1970).

120. A. Boudin, S. Deutsch, F. Hanappe, and M. Vosters, Proc. International *Symposium Applications of Neutron Activation Analysis in Oceanography*, Brussels, 1968, 13 (Inst. Roy Sci. Nat. de Belgique, 1968).

121. D. F. Schutz and K. K. Turekian, *Geochim. Cosmochim. Acta* **29**, 259 (1965).

122. D. E. Robertson, L. A. Rancitelli, and R. W. Perkins, *USAEC Report BNWL-SA*-1776 (1966).

123. D. E. Robertson and J. M. Prospero, *USAEC Report BNWL*-1051 (*part* 2) **53**, (1969).

124. A. A. Smales and B. D. Pate, *Analyst* 77 (1952).

125. A. I. Ryabinin and A. S. Romanov, *Geokhimiya* 875 (1970).

126. K. K. Bertine, *USAEC Report TID*-25642 (1970).

127. L. A. Rancitelli, W. L. Templeton, and J. M. Dean, *USAEC Report BNWL*-1051 (*part* 2) 152 (1969).

128. L. A. Rancitelli, *USAEC Report BNWL*-1051 (*part* 2) 146 (1969).

129. R. Fukai and W. W. Meinke, *Limnol. and Oceanogr.* **4**, 398 (1959).

130. H. Hamaguchi, R. Kuroda, and K. Hosohara, *J. Atom. Energy Soc. Japan* **2**, 317 (1960).

131. P. F. Wiggins, F. E. Senftle, and D. Duffey, *Trans. America Nucl. Soc.* **12**, 492 (1969).

132. M. Merlini, F. Girardi, and G. Pozzi, *Nuclear Activation Techniques in the Life Sciences*, Proc. Symposium, Amsterdam, 1967, 615 (International Atomic Energy Agency, Vienna, 1967).

133. M. Merlini, C. Bigliocca, A. Berg, and G. Pozze, *Nuclear Techniques in Environmental Pollution*, Proc. Symposium, Salzburg, Oct. 26–30, 1970, 447 (International Atomic Energy Agency, Vienna, 1971).

134. M. Merlini, O. Ravera, and C. Bigliocca, *N.B.S. Spec. Publ.* 312 vol. 1 475 (1969).

135. M. Merlini, in *Impingement of Man on the Oceans* (John Wiley, New York 1971) D. Hood, ed. Chap. 17.

136. G. W. Leddicotte and D. W. Moeller, *USAEC Report ORNL-CF* 61-5-118 (1961).

137. R. L. Blanchard, G. W. Leddicotte, and D. W. Moeller, *J. Am. Water Works Assoc.* **51,** 967 (1959).

138. W. H. Funk, S. K. Bhagat, and R. H. Filby, Proc. *Eutrophication-Biostimulation Assessment Workshop*, 1969, 207 (Univ. Calif. 1969).

139. L. A. Rancitelli and T. M. Tanner, *USAEC Report BNWL*-1051 (*part* 2) 137 (1969).

140. G. Schmidt, *Report KFK*-863 (1968).

141. D. P. Kharkar, D. D. Turekian, and K. K. Bertine, *Geochem. Cosmochim. Acta* **32,** 285 (1968).

142. O. Landström and C. G. Wenner, *Report AE*-204 (1965).

143. G. W. Leddicotte, *N.B.S. Spec. Publ.* 312 vol. 1 76 (1969).

144. R. Draskovic, T. Tasovac, and R. Radosavljevic, *Nuclear Techniques in Environmental Pollution*, Proc. Symposium, Salzburg, Oct. 26–30, 1970 329 (International Atomic Energy Agency, Vienna, 1971).

145. W. F. Merritt, Int. Symposium *Identification and Measurement of Environmental Pollution Abstracts*, June 14–17, 1971, Ottawa, No. 65A (1971).

146. D. McKown, M. Kay, D. Gray, A. Abu-Samra, M. Eichor, and J. Vogt, Proc. American Nuclear Society Topical Meeting *Nuclear Methods in Environmental Research*, Columbia, Mo., Aug. 23–24, 1971 150 (Univ. Missouri, 1971).

147. M. Wahlgren, F. R. Rawlins, and D. N. Edgington, Proc. American Nuclear Society Topical Meeting *Nuclear Methods in Environmental Research*, Columbia, Mo., Aug. 23–24, 1971 97 (Univ. Missouri, 1971).

148. K. Ljunggren, B. Sjöstrand, A. G. Johnels, M. Olsson, G. Otterlund, and T. Westermar, *Nuclear Techniques in Environmental Pollution*, Proc. Symposium, Salzburg, Oct. 26–30, 1970 373 (International Atomic Energy Agency, Vienna, 1971).

149. F. W. Lima and C. M. Silva, *J. Radioanal. Chem.* **1,** 147 (1968).

150. A. W. Gosling, E. A. Jenne, and T. T. Chao, *Econ. Geol.* **66,** 309 (1971).

151. P. Schiller and G. B. Cook, *Anal. Chim. Acta* **54,** 364 (1971).

152. K. D. Linstedt and P. Kruger, *Anal. Chem.* **42,** 113 (1970).

153. A. G. Souliottis, E. P. Belkos, and A. P. Grimanis, *Analyst* **92,** 300 (1967).

154. R. A. Wallage, W. Fulkerson, W. D. Schults, and W. S. Lyon, *Report ORNL-NSF-EP*-1 (1971).

155. N. Nelson *et al.*, *Environ. Research* **4,** 1 (1971).

156. L. Friberg, G. Lindstedt, G. Nordberg, C. Ramel, S. Skerfving, and J. Vostal, *E.P.A. Report APTD*-0838 (NTIS PB-205-000, U.S. Dept. of Commerce, Wash., D.C., 1971).

157. M. R. Greenwood and T. W. Clarkson, *Amer. Ind. Hyg. Assoc. J.* **31,** 250 (1970).

158. R. Christell, L. G. Erwall, K. Ljunggren, B. Sjöstrand, and T. Westermark, Proc. 1965 Int. Conf. *Modern Trends in Activation Analysis*, College Station, Texas, Apr. 19–22, 1965 (Texas A & M Univ., College Station, 1965).

159. R. H. Filby, A. I. Davis, K. R. Shah, and W. A. Haller, *Mikrochim. Acta* 1970 1130 (1970).

160. B. Sjöstrand, *Anal. Chem.* **36,** 814 (1964).

161. L. Kosta and A. R. Byrne, *Talanta* **16,** 1297 (1969).

162. J. M. Rottschafer, J. D. Jones, and H. B. Mark, Jr., *Environ. Sci. Technol.* **5,** 336 (1971).

163. J. D. Jones, J. M. Rottschafer, H. B. Mark, Jr., K. E. Paulsen, and G. D. Patriasche, *Mikrochim. Acta* 1971 399 (1971).

164. G. V. S. Rayudu, B. Tiefenbach, and R. E. Jervis, *Trans. Amer. Nucl. Soc.* **11,** 54 (1968).

165. V. P. Guinn and R. Kishore, Proc. American Nuclear Society Topical Meeting *Nuclear Methods in Environmental Research*, Columbia, Mo., Aug. 23–24, 1971 201 (Univ. Missouri, 1971).

166. G. E. Miller, P. M. Grant, R. Kishore, F. J. Steinkruger, F. S. Rowland, and V. P. Guinn, *Science* **175,** 1121 (1972).

167. J. I. Kim and H. Staerk, *Radiochim. Acta* **13,** 213 (1970).

168. G. Henke, S. Westerboer and H. Portheine, *Arch. Toxikol.* **23,** 293 (1968).

169. E. J. Kennedy, R. R. Ruch, and N. F. Shimp, *Environ. Geol. Notes* No. 44 (1971).

170. A. R. Byrne, M. Dermelj, and L. Kosta, *Nuclear Techniques in Environmental Pollution*, Proc. Symposium, Salzburg, 26–30 Oct. 1970 415 (International Atomic Energy Agency, Vienna, 1971).